MW00711082

Publisher
Loft Publications

Editorial Coordinator
Joaquim Ballarín i Bargalló

Documentation
Mariona Villavieja i García
Joaquim Ballarín i Bargalló

Texts
Maria Cinta Martí i Amela
Joaquim Ballarín i Bargalló

Graphic design
Arguelles Gutierrez Disseny / Juan Prieto

Layout
Anna Soler i Feliu / Juan Prieto

Cover layout
María Eugenia Castell Carballo

Loft Publications, S.L.
Via Laietana 32, 4º, Of. 92
08003 Barcelona, Spain
P +34 932 688 088
F +34 932 687 073
loft@loftpublications.com
www.loftpublications.com

ISBN: 978-84-96936-97-3

Printed in Spain

Contents

INTRODUCTION

Wood, used since the dawn of time as a building material, brings together a wide range of characteristics that endow it with great versatility. It is an eco-friendly structural material with thermal and acoustic insulation properties and almost unlimited applications in interior design and decoration, not to mention more subjective aspects, such as warmth, texture, and touch.

Today, wood or timber is considered a raw material that is inexhaustible and consistent with the concept of sustainability. The use of information technology in its handling has resulted in the systematization of the cut. The appearance of a multitude of derived products has optimized the exploitation of otherwise non-usable leftover material. There has been an advance in the design and operation of joining and mounting systems, panel-based construction, reliable renovation systems, maintenance systems, and biocide systems against pests for the purposes of preservation and upkeep.

Because of these advances, wood forms an integral part of prefabricated construction, largely manufactured in a workshop, with resulting time savings and the possibility of using advanced machinery. This results in a light, transportable, easy-to-use product, which is assembled at the final stage without the strict need for a specialist workforce. This material promotes energy savings during handling and installation. In short, it is 100% recyclable.

TIMBER HOUSE

The earliest construction with wood took place in Northern Europe, reaching the rest of the world with settlers from this region, whenever and wherever suitable quantities of quality wood were to be found. At first, they were built simply from logs with the bark planed down before being piled up horizontally and jointed in the corners. Later, the arrival of sawmills led to their measurements being optimized and standardized.

This construction system is currently marketed in "kits", which are completely made up in the workshop and rapidly assembled at their final destination. Logs offer many possibilities for the finish: round, sawn, dyed, weathered, or painted..., as well as designs that are much less straight up and down as they were in the very beginning.

HEAVY FRAMING

Developed in Europe and China in the Neolithic period, heavy framing spread to North America, Japan, and Southeast Asia, reaching its greatest splendor between the end of the Middle Ages and the 19th century. The modern version of this system made its appearance in Europe at the end of the 1960s and Japan at the beginning of the 1980s, with the use of new steel joining systems, increasing the resistance, prefabrication, and systematization of the process.

Heavy framing is based on a main structure of pillars and large-span girders, which are normally in modules, supporting a lesser structure made of joists. The whole system should be correctly braced.

LIGHT FRAMING

Heavy framing developed into this new system in the United States during the 19th century. Light framing does not differentiate between load and wall elements; instead, it acts as a spatial structure, with load-bearing planes in the three main directions built from small components and using very simple joints, fasted with nails or staples. It is a standardized and modular system, albeit with considerable flexibility, and can be given a large degree of prefabrication.

The development of wood-based boards began towards the second half of the 20th century. Its most emblematic representative is the plywood board. Its evolution is closely linked to that of adhesives, and it revolutionized the structural system of the light framework.

ENGINEERED WOOD

Engineered wood, in the form of glued laminated timber, appeared at the beginning of the 20th century for straight, large-span beams. Curved beam systems appeared later. The system is based on techniques similar to those of light framing; it uses small pieces of timber to produce pieces of almost unlimited size. In the wake of the 1910 International Exhibition of Brussels, where the product was unveiled, its use enjoyed a high level of acceptance in Europe and especially Switzerland.

The system spread to areas outside of Europe after 1923. At the end of the 1930s, when its worthiness as a construction system was verified and materials like steel suffered the consequences of restrictions caused by the Second World War, its use was established definitively. Coinciding with this advance, the invention of synthetic adhesives enabled the use of engineered wood in virtually any situation, without the limitations of adhesives for indoor use.

01

wood
in the landscape

Wood integrates well with its surroundings and has a very wide range of applications. Use in landscaping is just another of the possibilities it offers. The projects described in this chapter become the feature of their setting, communicating with it on equal terms, creating their own interior gardens, and becoming an integral, almost necessary, part of the landscape.

MAISON CONVERCEY

capturing the landscape

QUIROT-VICHARD ARCHITECTES
COLLABORATORS: Alexandre LENOBLE, Emmanuel BEAUDOIN
PHOTOS: © Luc BOEGLY

GRACHAUX – FRANCE

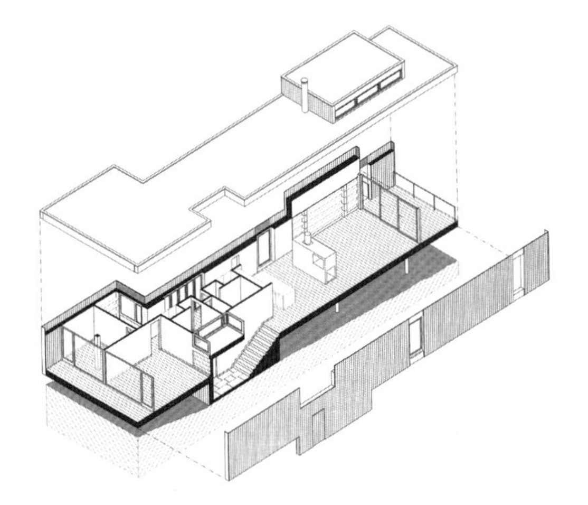

Raised above the ground on pillars, the building floats lightly over the expanse of the landscape. From the interior, everything focuses on achieving a perfect view of the exterior through a frame. The building, which is narrow and stretched in both directions, takes on the appearance of a kaleidoscope, distilling the plain into many different images.

A narrow steel structure, erected in a couple of weeks, received a warm timber façade, meaning that a supposedly austere structure integrates itself perfectly into the environment. The only point of contact with the ground, the entrance, acts as a balancing and anchorage point for this tetrahedron as it sails on a green sea.

Slight height variations in the height of the rooms, constantly controlled openings, and the continuous dialog between interior and exterior enrich a design which consciously minimizes its impact on the area. Inside, the spatial continuum is only broken to give a unique feel to the eating and living spaces.

LAROUSSE

MAISON GOULET

welcoming retreat

SAÏA BARBARESE TOPOUZANOV ARCHITECTES
PHOTOS: © Marc CRAMER, Frédéric SAIA

SAINTE MARGUERITE DU LAC MASSON - CANADA

Standing alone in the midst of a rugged landscape, this house is the very embodiment of a retreat. Set on a north-south slope facing a lake, the building is located on a rocky expanse, stretching from east to west. Two large stone chimneys give the appearance of anchoring the wooden structure to the ground. Zinc cladding lends it a velvety, almost cold, texture, contrasting with the warmth emanating from its interior.

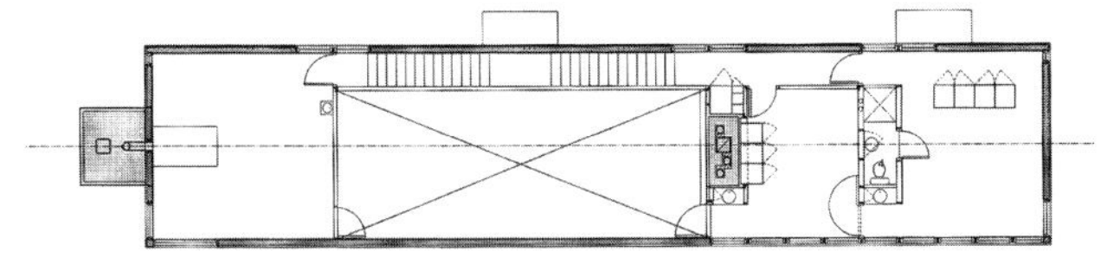

When the weather allows, the large glazed doors and walls open to dissolve the barriers and enable the exterior spaces to be an extension of the interior. On the uppermost level, the windows on both sides of the building interact effortlessly, allowing the occupants to look out over the surrounding area and appreciate their location on the slope.

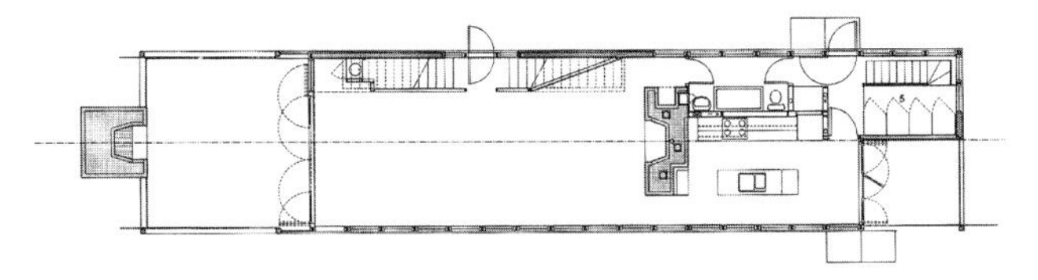

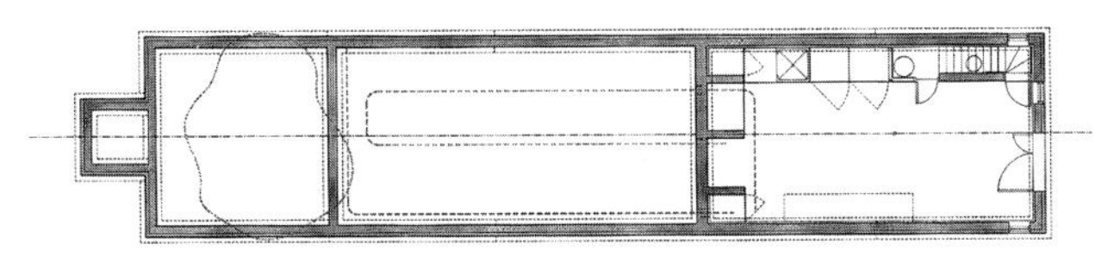

As architecture deeply rooted into the topography of the site, its climate, and its characteristics, the house remains open to interpretation. The volume is, in principle, traditional and free from superfluous details, proposing a deeper relationship with an environment it protects from, while it avoids the gray areas of nostalgia and convention.

CHUN RESIDENCE

levels of introspection

CHUN STUDIO, INC.
COLLABORATORS: Jamie BUSH, Jon BRODY
PHOTOS: © Tim STREET-PORTER

SANTA MONICA - CALIFORNIA - USA

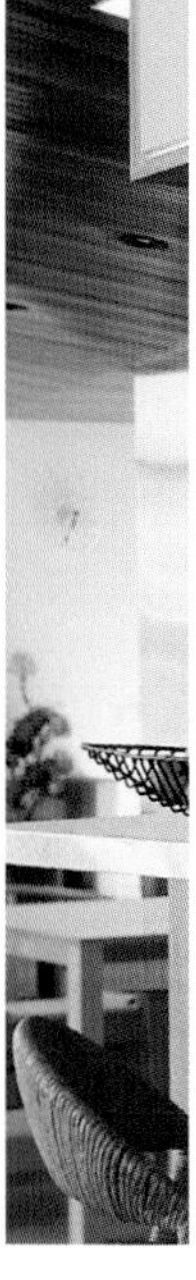

This H-shaped residence has a two-story street frontage acting like a great wall, concealing the occupants from passers by, but dissolving for visitors by means of a double height void at its core: a transition point between the opacity of the façade, asserted through the use of stone, and the transparency of the interior spaces, a vision in glass and wood.

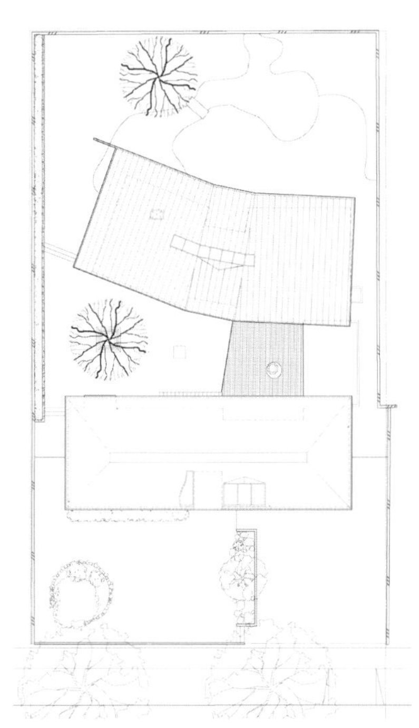

The idea behind the project was to create an independent, manageable space removed from the outside world. Visitors' perceptions are transformed as they go from the sheer size of the exterior walls to the human scale of the interior living spaces and gardens. A second contrasting feature comes in the form of the dialog between the glazed interiors and the gardens, particulary evident in the Japanese garden.

242

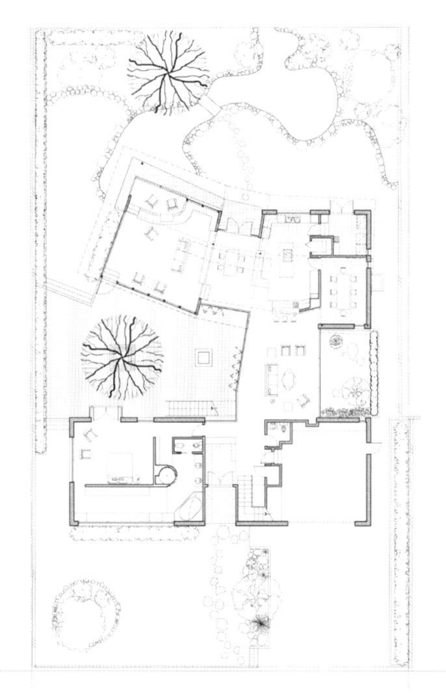

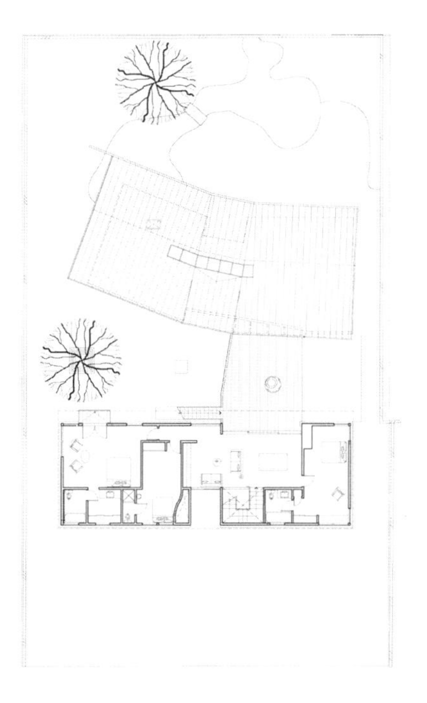

The interior opens up completely to visitors, with the kitchen acting as a focal point for the rest of the rooms. Interior façades are expanses of glass, creating a space that expands progressively, first inside the house and then towards the garden, stimulating states of mind that evoke weightlessness, peace, and silence.

ISLAND HOUSE

the boundaries of nature

ARKITEKTSTUDIO WIDJEDAL RACKI BERGERHOFF
PHOTOS: © Ake ERIKSSON-LINDMAN

STOCKHOLM – SWEDEN

The home is set on a coastal property located on one of the islands in the Stockholm Archipelago. Oaks and pines survive in the little fertile soil to be found between the sculptured rocks that are the predominant feature of the landscape. An oak believed to be around 500 years old dominates the raised area of the site.

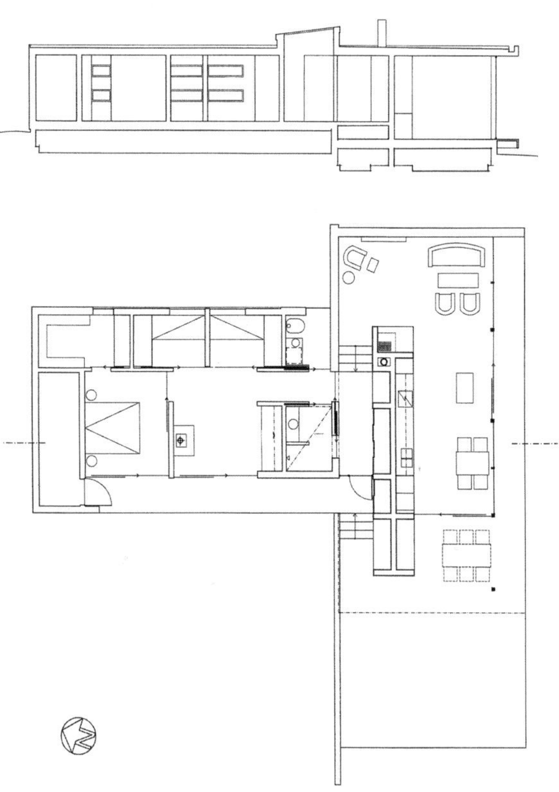

The clients, a young family with two children, wanted a home that took advantage of the magnificent views offered by the site, but without becoming an overpowering feature of the coastline. The owners also wanted to build a vacation home inspiring comfort and relaxation, a place for socializing and welcoming friends.

The building is set back from the coastline and slightly raised, half-hidden among the tree trunks, but faces the ocean through an open terraced area. It consists of a platform floor, a flat roof, and a series of shared spaces, the majority of which open by means of sliding doors. The most private areas of the house are located at the back, looking out over the forest, among rocks and an ancient oak.

02

changing wood

We usually consider buildings to be unchanging monoliths. The following projects show off the versatility of wood: transformed spaces, images of wood that varies with the passing of the hours, of the seasons, with the intensity of the sun, and with the occupation or use given to the home.

SUITCASE HOUSE HOTEL

versatile sensations

EDGE DESIGN INSTITUTE LTD.
GARY CHANG
COLLABORATORS: Andrew HOLT, Howard CHANG,
Popeye TSANG, Yee LEE

BEIJING – CHINA

This project came about through a competition for 12 young promises of Asian architecture, which involved designing 12 community dwellings at the foot of the Great Wall of China. The project consists of a volume with clean lines, that was light, designed with a layered interior, and which could adapt brilliantly to the different uses it had during the day. This space was intended for pleasure and leisure, and featured wood as its unifying material.

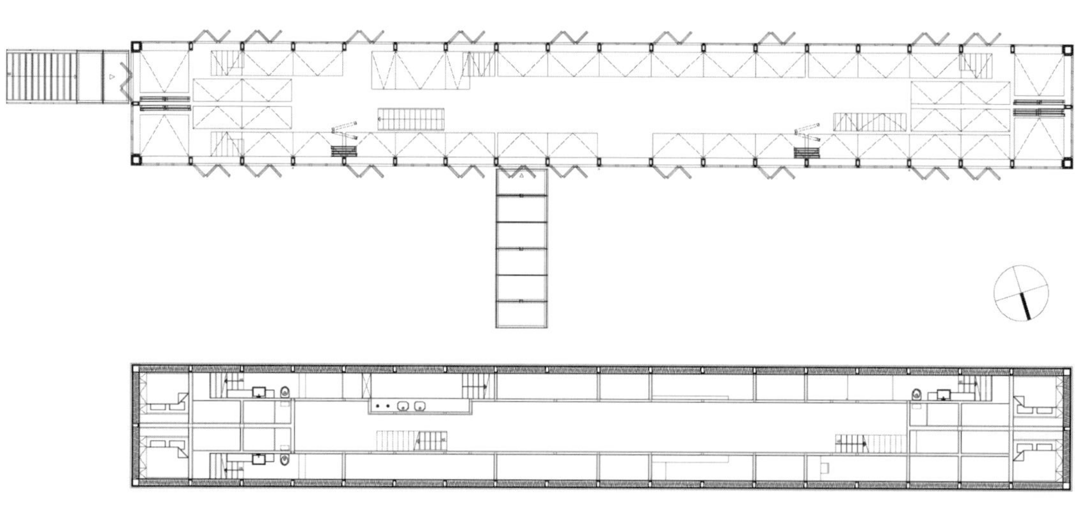

Designed on three levels, the interior layout is very adaptable to diverse uses – during the day, the upper level opens out completely, creating the image of a 150-ft long open pavilion. Later in the day, different spaces can be opened up (or closed, in fact) for leisure purposes: for listening to music, reading, meditating, bathing, and having a sauna, among others. The building can generate the relevant amount of sleeping spaces to accommodate more residents, or, alternatively, for parties.

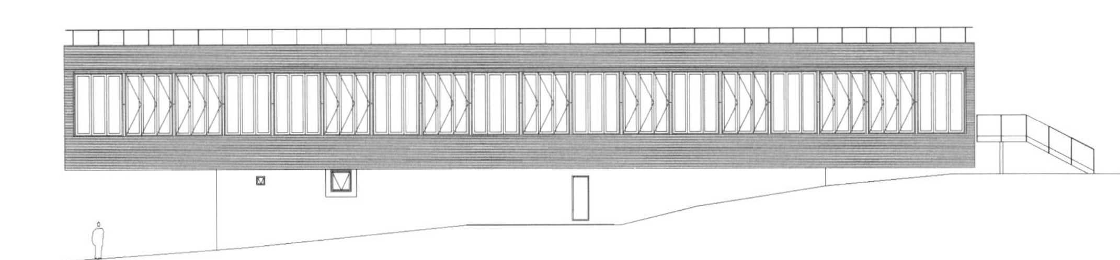

Reflecting its fluid internal layout, the façade comprises a repeated series of identical windows, like a stratified set of vertical layers, enabling the exterior to remain constant even when the internal layout varies. Wood covers the frame and all other structural elements of the residence, blurring the limits of the house, the interior, and the furnishings.

PUTNEY HOUSE

split personality

TONKIN ZULAIKHA GREER ARCHITECTS
PHOTOS: © Patrick BINGHAM HALL

SYDNEY – AUSTRALIA

Breaking with the passive uniformity of the surrounding urban setting, this home stands defiant, expressive, playful, restrained, and timeless. Situated right on the bank of the Parramatta River, where it becomes Sydney Harbour, the light from the north and the darkness from the south define the dual nature of its forms – both open and reclusive. It smiles brightly and cheerfully towards the north while politely greeting the south side with more restraint.

An enclosed courtyard on the north side, the focal point for the different living areas – the main one of double height, provides seclusion, while the south side features three pavilions that recall the ideal of the classical villa. A gallery bridge provides access to the upper floor, housing the main bedrooms and the study.

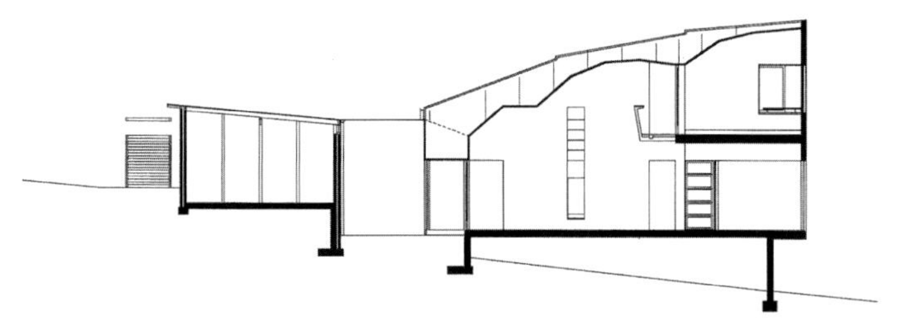

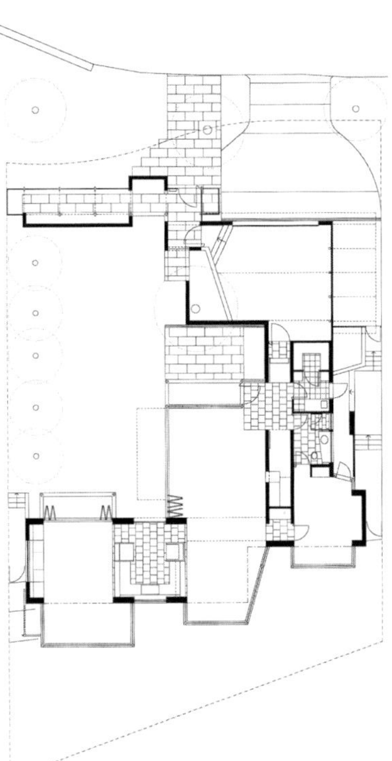

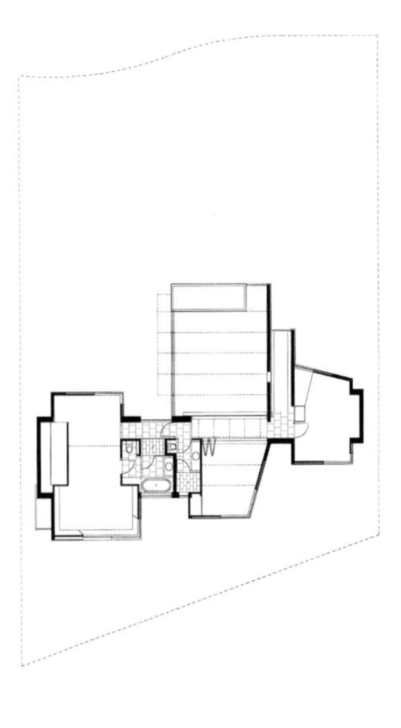

The triangles formed by the north-facing roof contrast with the more conventional horizontal straight line of the south façade. The differently-angled planes formed by the wood-lined ceiling contrast with the sky outside and provide a cooling effect for the house. Frameless glass expanses take the place of walls, letting the exterior landscapes penetrate into the house.

MAISON PLÉNEUF-VAL-ANDRÉ

between presence and absence

DL ARCHITECTS
COLLABORATORS: Eco-Bois, Gildes DARRAS
PHOTOS: © Stephen LUCAS

ITÚ – FRANCE

This building was constructed in two successive stages. The future guest accommodation was built first, according to a program that was reduced but sufficient. The second stage involved the main residence, and consisted of a more substantial program. Finally, the two buildings, actually a single unit divided into two to create a sense of mutual independence, were joined by a terrace built as a platform over the ground.

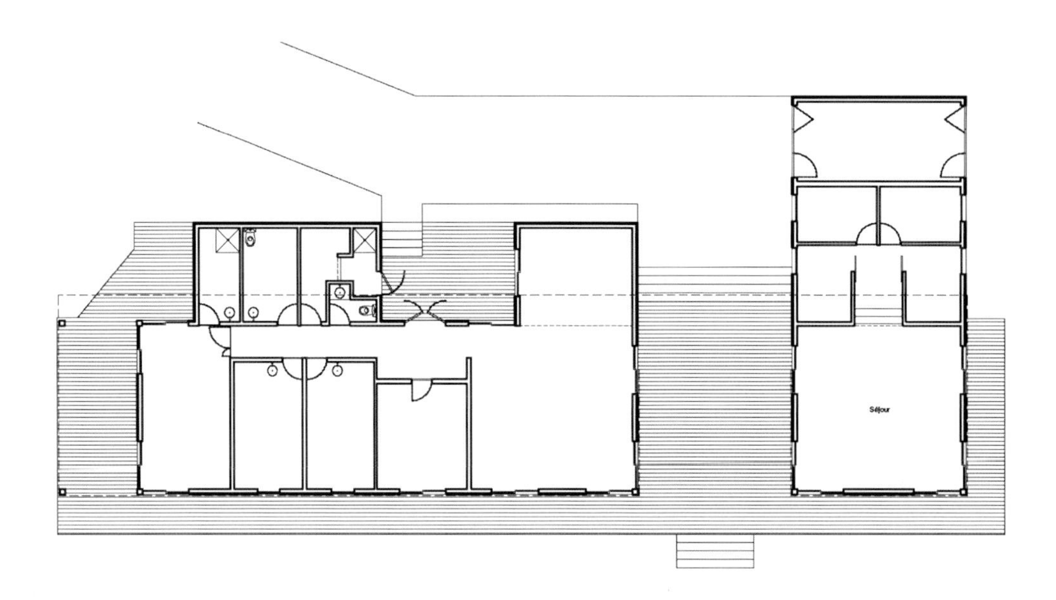

A residence used only on certain occasions presents an obvious contradiction; when it is occupied, it will have to open up to the landscape, allow the sun to enter, and provide access to the exterior spaces. On the contrary, when not in use, it will have to avoid showing signs of being unoccupied and guard itself from the same exterior it had previously opened itself up to. The design overcomes this challenge by incorporating mobile façades that open and close, alternately giving the impression of an accessible home and an impenetrable monolithic mass.

Construction deadlines and a desire to respect the nature of the building's setting steered the designers towards the choice of wood as a building material. They used kerto, fir, and cumaru to produce a very unified result. The rooms are laid out according to the aspect of the house; the service spaces (kitchen and bathrooms) face north, while the bedrooms, living areas, and terraces face south, enabling the light to penetrate through its large glazed expanses and reach the back of the house.

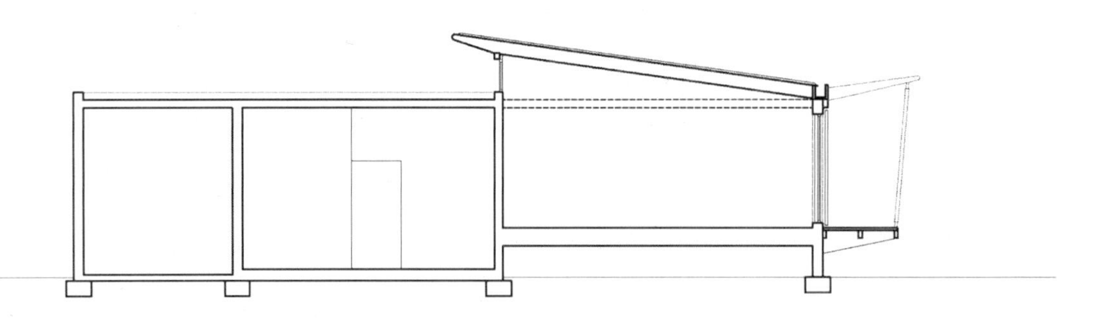

HAUS NENNING

contextual and adaptable

CUKROWICZ-NACHBAUR ARCHITEKTEN
Andreas CUKROWICZ, Anton NACHBAUR-STUM
COLLABORATORS: Christian MOONSBRUGGER, Saskia JÄGER, Marcus CUKROWICZ
PHOTOS: © Hanspeter Schiess Fotografie

HITTISAU – SWTIZERLAND

This free-standing structure was built by and for a carpenter, and is located on the main street of a small town, next to the church. The layout consists of three levels. The lowest contains a small shop, communal areas, and the entrance; the other two contain the south-facing family home, separated by the stairwell from an independent north-facing apartment, which can be integrated to enable the different generations of the family to mix.

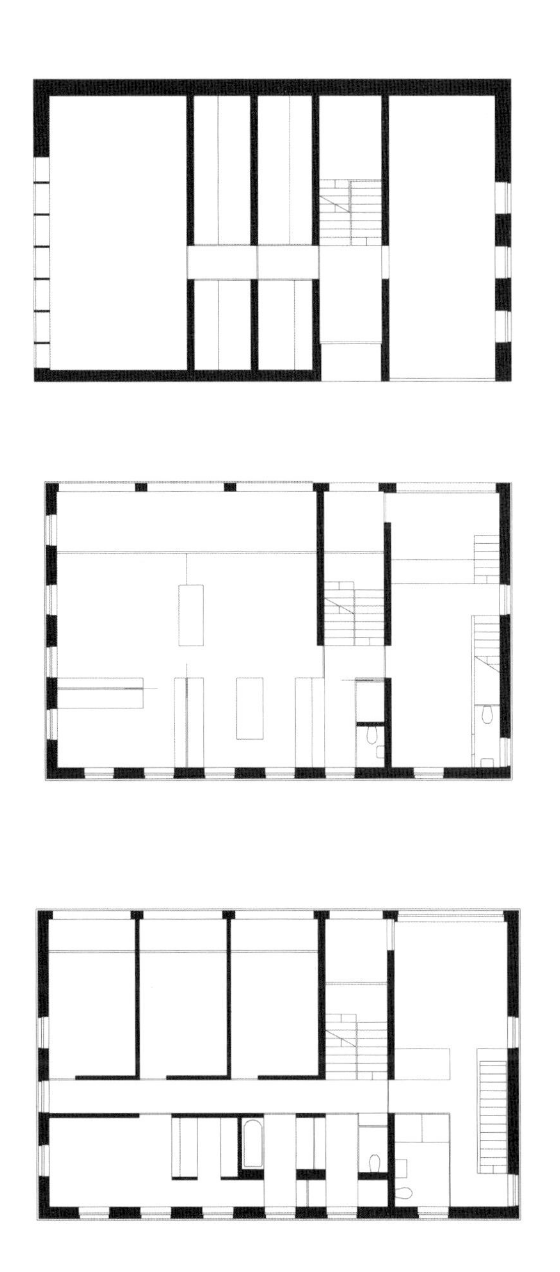

The slight surface variations on each floor and the sliding window blinds protect the façade against the elements. Simple compositional features and the exclusive use of solid timber as the building material create the style of the house. The amount of light entering the windows can easily be adjusted by sliding the blinds.

This three-floor timber structure reproduces the typical motifs and elements of its surroundings and interprets them in a new context. The result is a house that does not appear out of tune with its environment, despite creating its own language. Perfectly integrated, this house was not intended to dazzle; it only attracts the attention of passers-by when they give it a second glance.

03

expert wood

On occasions, designers team up professionally with carpenters. This collaboration gives rise to projects that treat space and the material with the same sensitivity and knowledge. These four projects, located in very different places, reflect this treatment in their implementation, through innovating, renovating, systematizing, or simply using wood with great precision.

CASA DMAC

a feel for the material

NASSAR ARQUITECTES, S.L.
Daniel NASSAR
COLLABORATORS: Domingo GONZÁLEZ,
Joaquim BALLARÍN
PHOTOS: © Daniel NASSAR

VALLROMANES – SPAIN

Designed with a marked linear layout, this home divides the site, opening out fully towards the south, source of the views and light, while turning its back on the north, where the entrance is located. The building takes advantage of a natural incline to provide access to the upper level, which is set aside for living areas, whereas the lower level, housing the more private spaces, provides more direct contact between the rooms and the terrain.

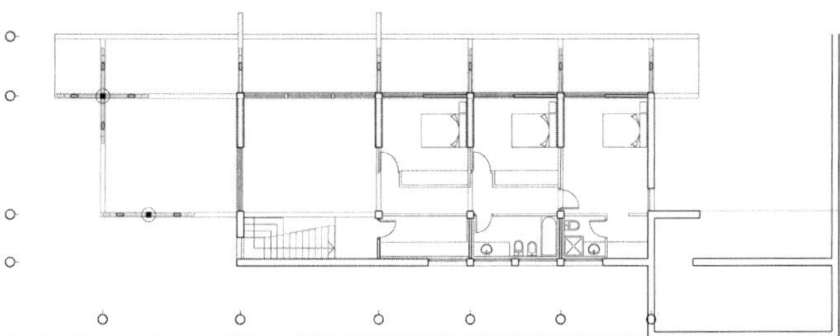

The terrace recalls the deck of a ship. The positioning of floor boards and the slats on the wall, the lines of which fade into the distance, counterbalance the roof beams. The module containing the living module is marked out on the façade as it breaks up the continuity of the terrace and the roof. The entrance to the house is parallel to a textured stone wall, which separates it from the exterior while hinting at the openness that lies behind the wall.

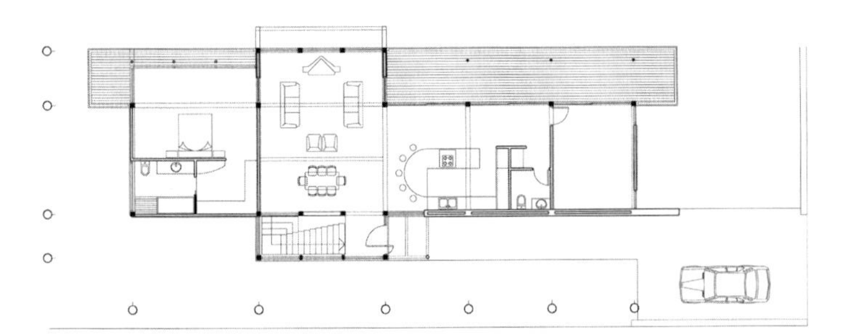

The continuity of the space is accentuated by not letting the partiution walls reach the ceiling. This enables the roof frame to be seen from adjoining spaces and the careful positioning of the beams and the almost ironic use of metal struts to be appreciated. The staircase is a metal structure painted white and lined with solid wood, set neatly against a white brick wall to connect its structural function with the wood, which has an enamel finish in the same color.

DANIELSON HOUSE

maritime construction

BRIAN MACKAY–LYONS
COLLABORATORS: Trevor DAVIES, Bruno WEBER, Darryl JONAS
PHOTOS: © Undine PRÖHL

SMELT BROOK – CANADA

This cliff top house was built for a meteorologist and a landscape architect with a modest budget and a set of environmental certifications. Its remote location and the clarity of the architectural program represent attitudes and a lifestyle that are increasingly becoming the most popular response to the everyday insanity pervading our society.

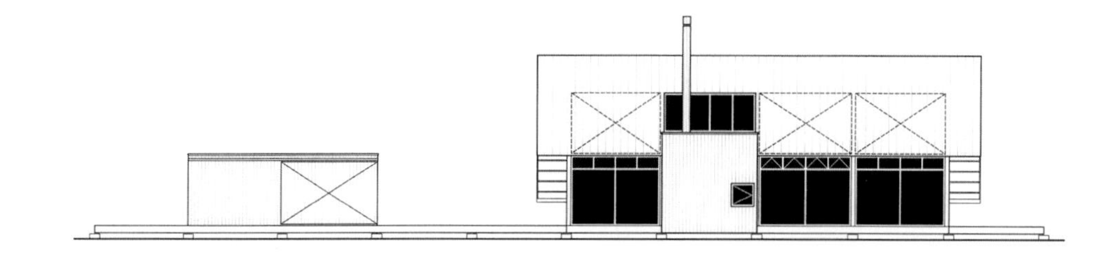

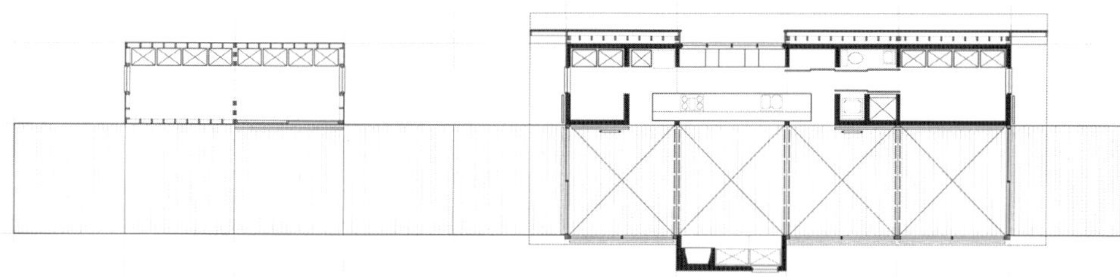

The program was designed as two structures built on a wooden platform, reflecting the horizontality of the sea and featuring a strict separation between the service spaces and the spaces they serve. The former are concentrated in an area on one side, freeing up an expanse for the latter, which can open out to the exterior when the weather permits.

To optimize the available budget, prefabrication and assembly took place in a workshop, using glued laminated timber to obtain squares of a sufficient size. This resulted in a system of beams and wooden pillars which was then given aesthetic expression by means of differently-textured claddings. The result is a building resembling a ship: a light and mobile structure which is equally at home on land, water, or ice.

MONTVIEW RESIDENCE

environment regained

JMA ARCHITECTS
John MAINWARING
COLLABORATORS: Jo CASE, Steve GUTHRIE

KENILWORTH – AUSTRALIA

Mountview is the reinterpretation of a past way of life and the paradigm of a landscape that refuses to disappear. The project brought together the construction and remodeling of several buildings to endow the structure with an image similar to that of a rural settlement, where the different components are located on the site in no particular order, but respecting the trees and the topography, and becoming part of the landscape.

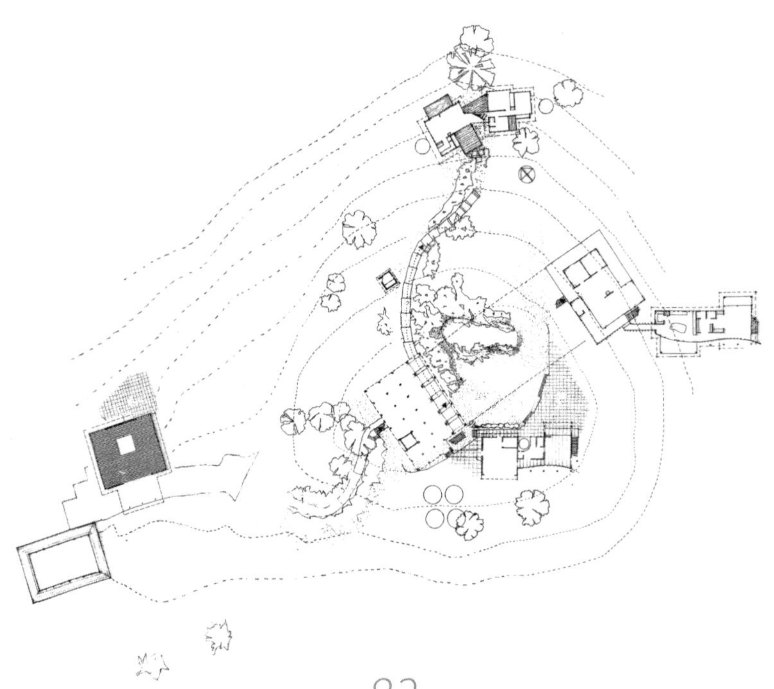

Termites had destroyed a large proportion of the original timber building dating from 1890. The refurbishment involved little change to the exterior aspect of the house, while the interior transformations are very evident. The majority of the partition walls were replaced with sliding slat walls and the finishes were improved.

The openings on both façades, the floor-to-ceiling windows, and the timber slat walls were designed with the aim of providing ventilation in the summer and turning the house into a greenhouse in the winter. Following the refurbishment, the interiors became extroverted, airy, and open to the views of the surrounding landscape. This was a dramatic change from the original introversion designed to protect the house from the outside.

PILOTPROJECT KFN

timber construction system

Johannes KAUFMANN, Oskar Leo KAUFMANN
PHOTOS: © Ignacio MARTÍNEZ

ANDELSBUCH – AUSTRIA

This home for two families is actually the pilot house for the KFN system, developed from a collaboration between the architects and a carpentry workshop with the aim of building modular homes: a 16.5 × 16.5-ft timber structure housing freely laid-out prefabricated spaces together with wet areas – kitchen and bathroom – and a variable system of up to 24 modules with a 16.5 × 9-ft façade.

The design and the project are developed in the architects' office, before being sent directly to the carpentry workshop. Here the different sections of the house are manufactured while the foundations are being laid on the site. The kit is finally taken to the site and assembled as if it were a doll's house. This process guarantees reduced building times and streamlined budgets, and the result is both attractive and comfortable.

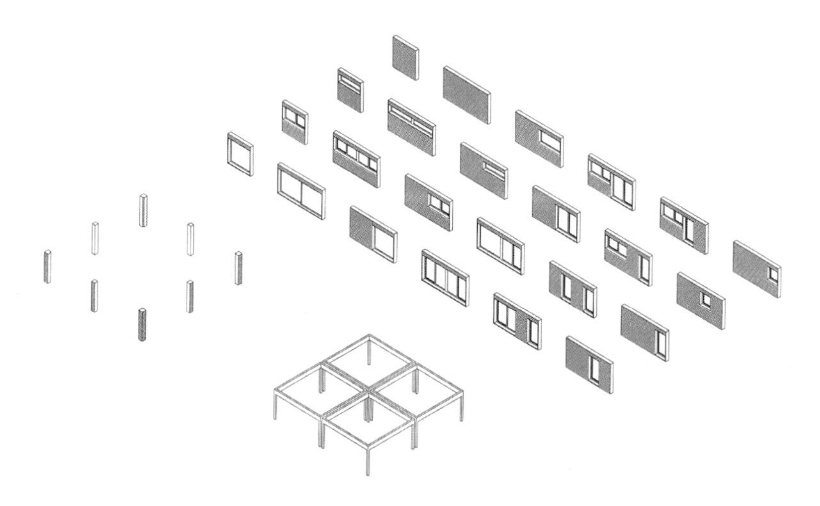

The framework of the house is made of red fir, while the rest of the façade and interior modules are compatible with any finish, preferably wood paneling or plywood. The system also incorporates various sustainability measures, integrating radiant floor heating with a recycled wood-fired hot water heater. The simplicity of building with this system recalls that of log cabins, but with a greater attention to design.

04

wood and nature

Unlike the vast majority of contemporary constructions, building with timber makes it possible for the structure to coexist with nature, where there are no winners or losers. The four following projects are defined by their environment; they show it great respect instead of destroying it with their presence.

CHILMARK RESIDENCE

a clearing in the forest

CHARLES ROSE ARCHITECTS

PHOTOS: © Chuck Choi Architectural Photography

CHILMARK – MASSACHUSETTS – USA

The site proposed by the owner posed a great challenge. This small clearing in the woods forced the designers of the building to break it up, adapting it to the scarce available building space in the thick vegetation. The original surroundings were given additional landscaping and drastic measures were applied during the construction to guarantee the protection and vitality of the natural environment.

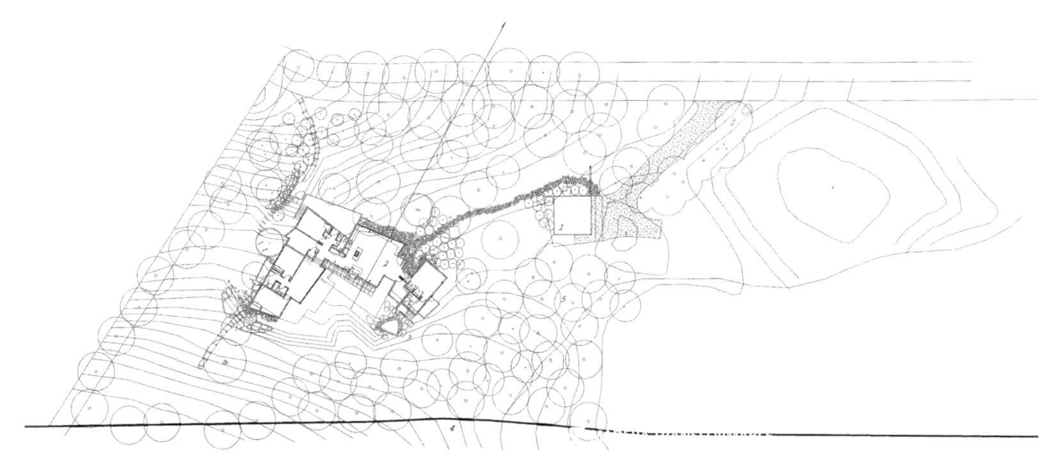

The dwelling is divided into three sections, with a large living area connecting the main bedroom to the guest wing, which can be closed in the winter. The central section combines the living, dining, and kitchen areas into a single, light, and modern space, which achieves a feeling of warmth through the use of cedar wood.

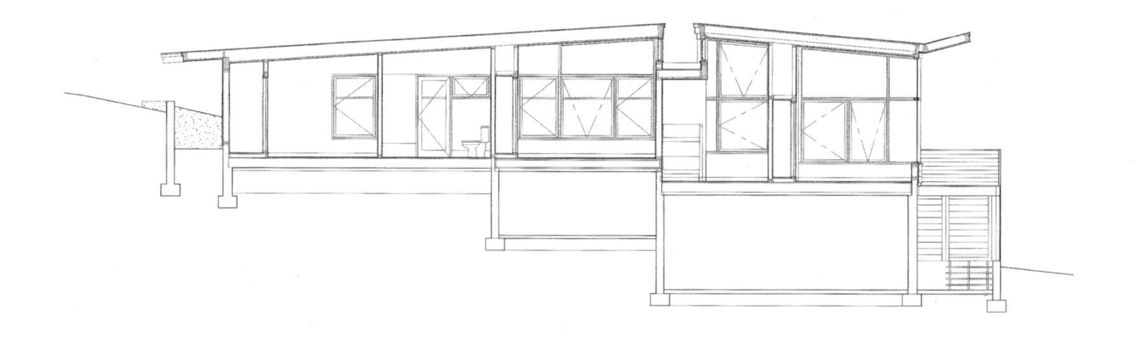

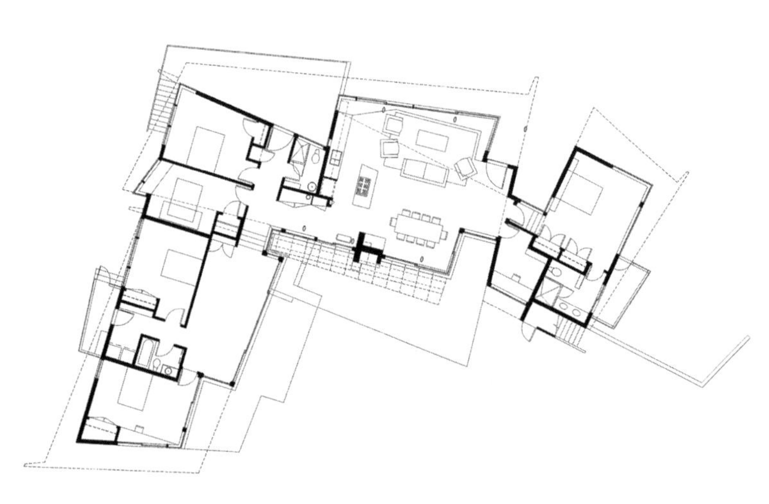

Elliptical pillars and large sliding doors make partition walls unnecessary in the interior. The living space is connected to the wooded landscape and the ocean beyond for the occupants' enjoyment. The decks and roof amplify the feeling that there are no barriers between interior and exterior.

VILLA LÅNGBO

the boundaries of nature

OLAVI KOPONEN ARCHITECT
COLLABORATORS: Oscari LAUKKANEN
PHOTOS: © Jussi TIANIEN

LÅNGHOLMEN KEMIÖ – FINLAND

Located on the western tip of an island and exposed to the prevailing winds, the building stands on the edge of a forest, meaning that only parts of it can be seen from the sea, while its occupants have an unobstructed view of the same sea from any one of its rooms. The building was designed as a farm. Some of its residential space can be turned into storage or production areas.

Built with the aim of demarcating human presence in the area, the limits between the dwelling and its setting are blurred. The spaces are individually defined with respect to light, sensations, and relation to nature, and are divided into private and communal spaces, laid out separately between a raised platform floor and a roof. Despite the simplicity of the structure, there is complexity in the spaces, which results in variation under the one roof.

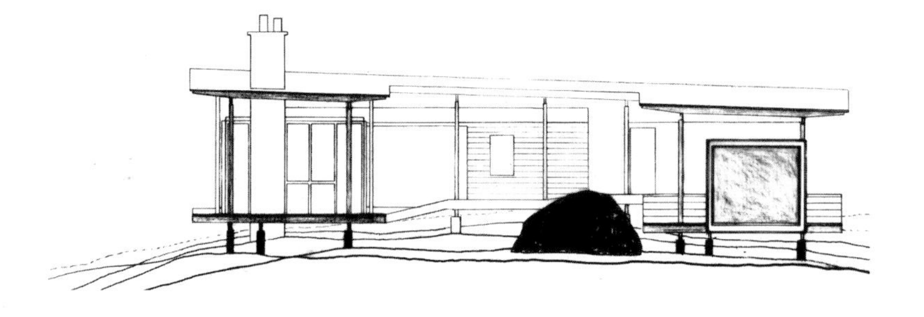

Exposed to the freezing and thawing of the sea, the house is not always accessible. It can only be reached by boat or skis, meaning that it cannot be used all year round. All of the materials are recyclable and the timber used was sourced locally. The house was built by hand. The materials were brought in by horse during the winter, and the environment was modified as little as possible.

KOEHLER RETREAT

eyrie

SALMELA ARCHITECTURE & DESIGN
David SALMELA, Souliyahn KEOBOUNPHENG
PHOTOS: © Peter BASTIANELLI-KERZE

SILVER BAY – MINNESOTA – USA

Distancing itself from modernity, but without turning its back on its concepts, this home is located on the edge of the wildest and most thrilling landscape. It is perched on the top of a cliff, casting a defiant look over Lake Superior.

The surrounding forest provided the cherry, maple, and cypress wood featured on the interior and exterior walls of the house. Windows, ceilings, stairs, and closets are made of maple, while the floors incorporate the textures of maple and cherry. The recycled cypress façade is a continuation of the natural landscape surrounding it, together with glued laminated timber post connecting the house to its roots in the solid rock.

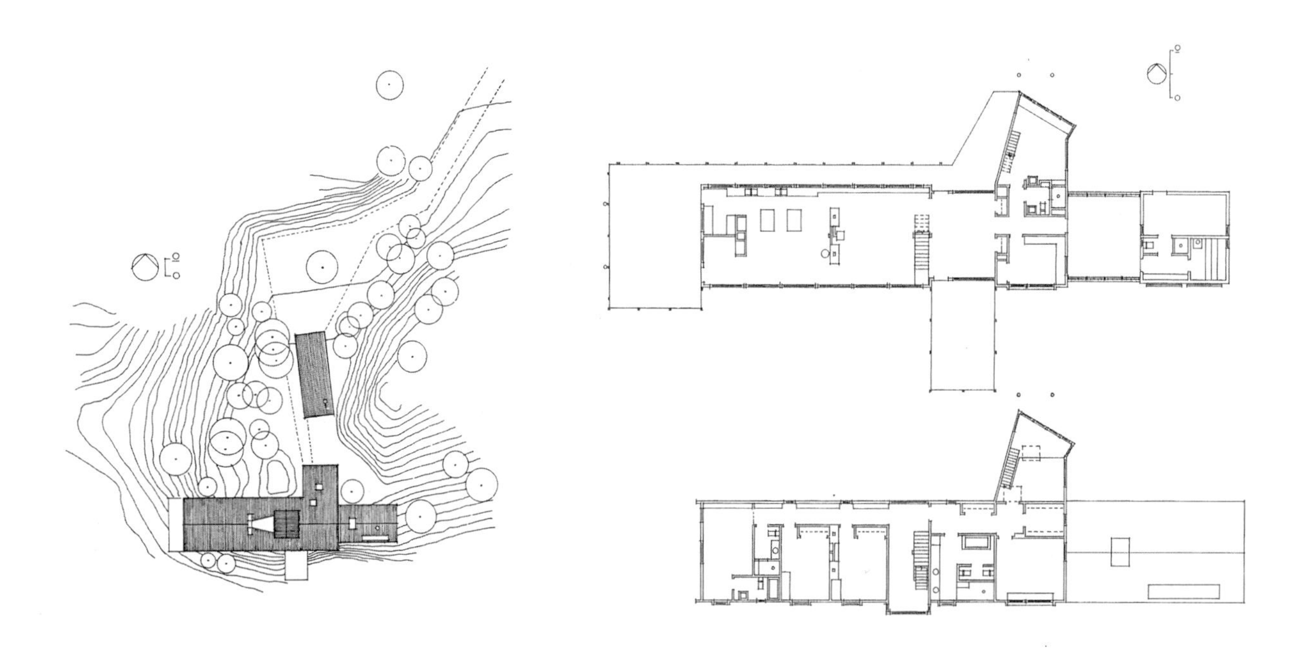

Its vacuous and unadorned interior awaits the arrival of the sunlight and the trees' shadows. The most inhabited areas of the house receive direct sunlight, while the study is left more subtle and filtered light that is more appropriate for the artist's work. The textures of the wood and the contrasting effect of light and dark as the day passes reflect the beauty of this retreat in the middle of nature.

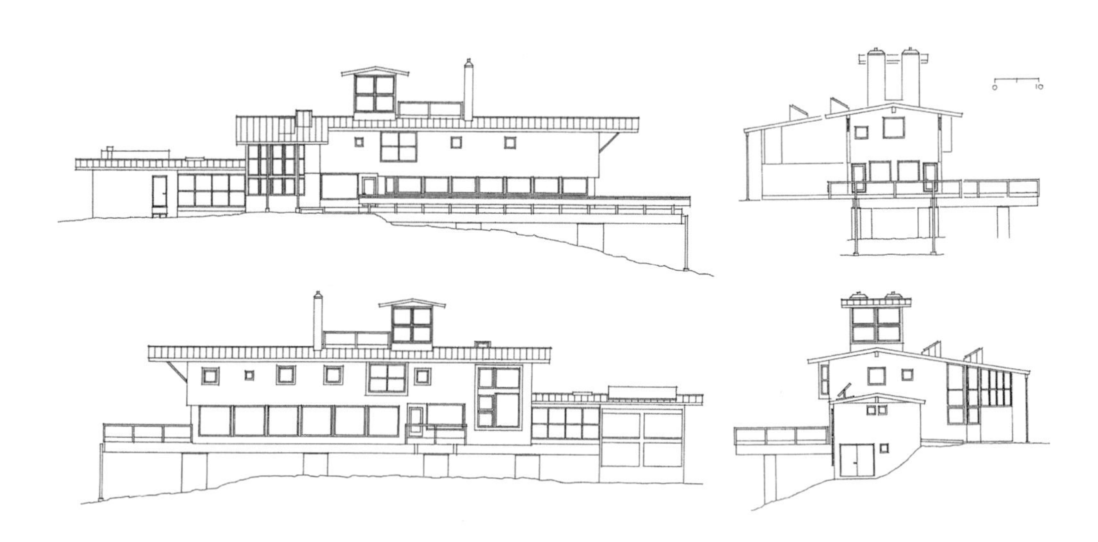

CASA ACAYABA

life among the trees

MARCOS ACAYABA ARQUITETOS
COLLABORATORS: Mauro HALLULI, Suely MIZOBE, Fabio VALENTIM
PHOTOS: © Nelson KON

GUARAJÁ – BRAZIL

This house is a model of how to build on a slope without damaging the environment. It is located 500 ft from the beach and 250 ft above sea level. An expanse of Atlantic Forest, the site was preserved by building the house on only three pillars placed among the trees, giving the house a hexagonal shape. This triangular structure, expanding outwards just like a tree, was assembled using beams and pillars in engineered jatobá timber, steel, cables, and braces.

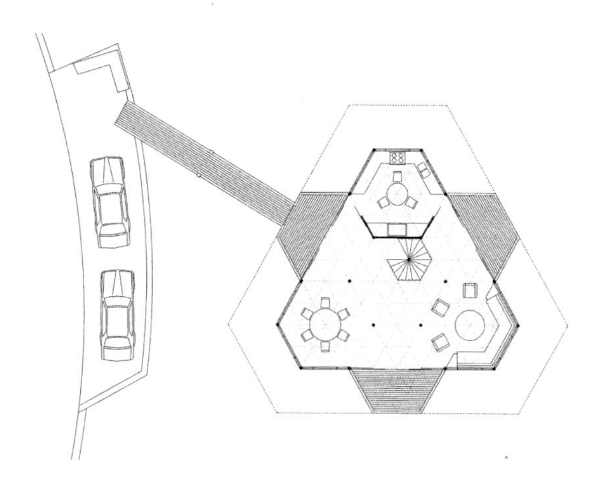

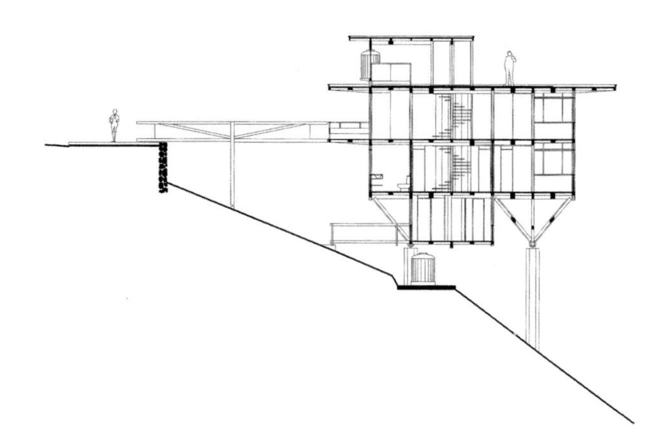

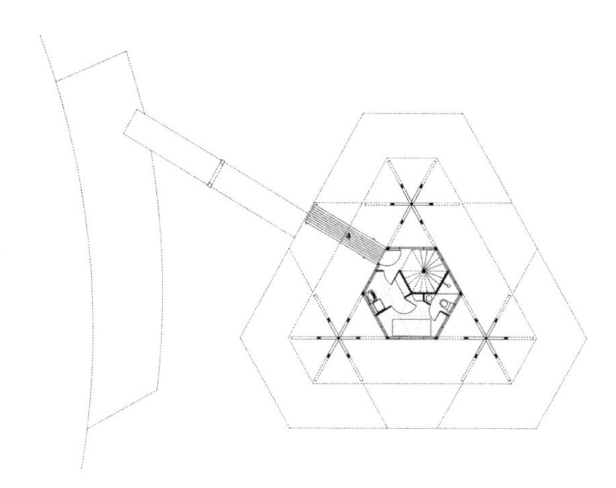

The roof is cantilevered out 7 ft to protect against tropical rains, providing space for an open-air living room above the treetops with views of both the summit of the hill and the ocean. Access to the main level, containing the living areas and the kitchen, is by means of walkway that is separate from the building. The bedrooms are located above, while the service areas are on the lower level.

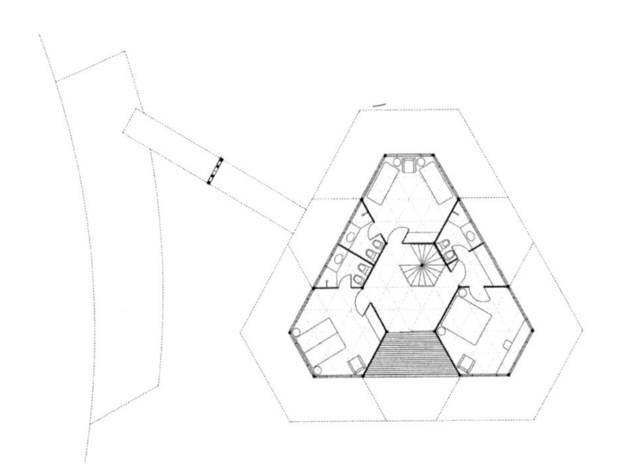

Designed for assembly using different lightweight pieces created in a workshop, the whole house was put together in four months by four people and with only the slightest environmental impact. The naturally rigid triangular form of the structure gives the spaces greater communication with the exterior. The façade gives the impression of being a succession of continuous balconies.

05

wood
in the flesh

Sometimes, there is no call for wood's load-bearing or other physical qualities. The look of wood alone when it is used wrapped around a building provides it with significant aesthetic qualities. Four projects from very different locations make use of this technique and provide a hint of the numerous possibilities wood offers in this field.

EAVE HOUSE

shelter from the storm

TEZUKA ARCHITECTS, MASAHIRO IKEDA

PHOTOS: © Katsuhisa KIDA

TOKYO – JAPAN

The monsoon rains that sweep the city of Tokyo for more than a month starting at the beginning of June define the character of this home. During this season of heat and heavy rains, the interiors of houses need to be able to withstand strong winds while being protected against the effects of water and humidity. A 16.5-ft cantilevered eave was designed to protect the open areas of the home from these weather events.

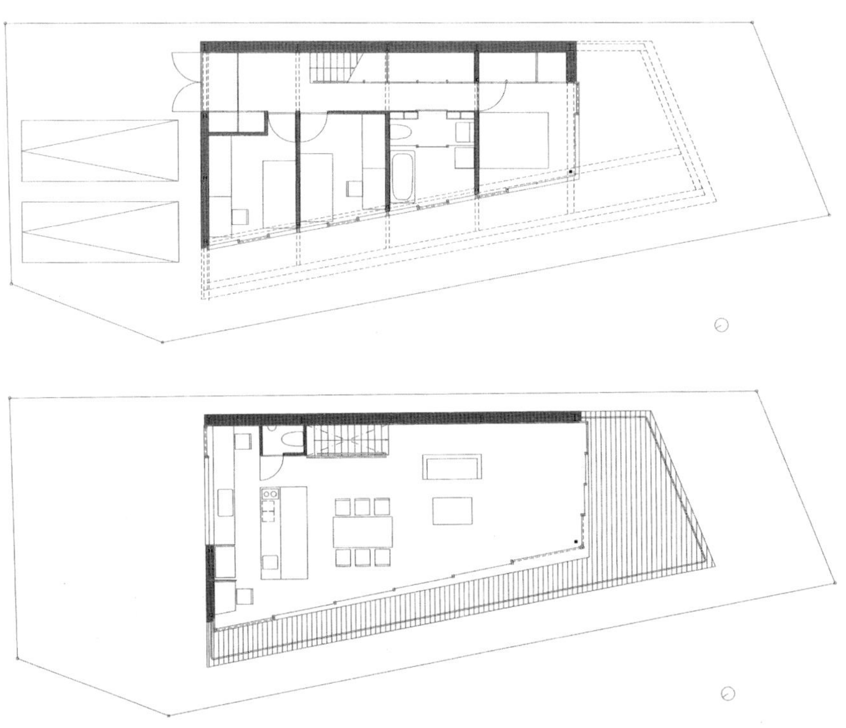

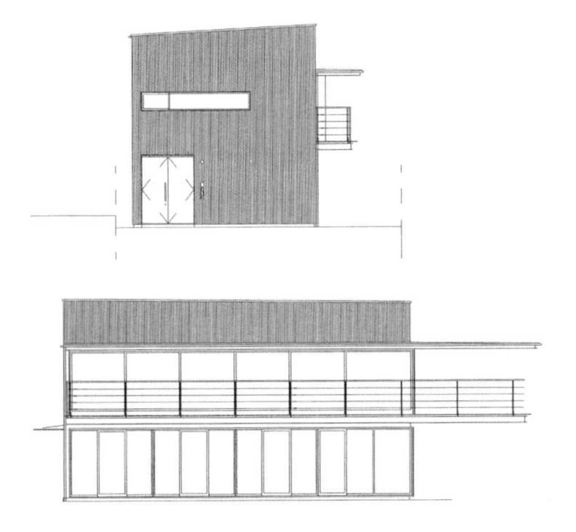

This project has a reduced and essentially domestic program that inverts the traditional layout. The most private areas – bedrooms and bathroom – are located on the lower level, leaving the upper level free for the living spaces, which enjoy excellent views as the result of the atypical layout. Both floors turn their backs on the party walls with adjoining constructions and open themselves up completely to the views.

The openings on the upper level façade are lower in height in order to give some control over the entering sunlight and protection from the rain. There is one solitary pillar in the most open corner of the façade, which blends with the sky when the sliding doors are opened. This action extends interior space out as far as the boundary of the eaves.

MEHRFAMILIENHAUS II

new to the neighborhood

VOGT ARCHITEKTEN
David VOGT
COLLABORATORS: Marcel KNOBLANCH, Antonella SILENO, Marc LIECHTI

MUHEN – SWITZERLAND

This apartment block makes its setting the main feature of its composition. Located in a small village surrounding a farmhouse dating from 1813, it uses contemporary language for integration and assimilation with the consolidated setting through the incorporation of local materials and the proportion of its volumes.

The building contains four apartments with two different layouts of great simplicity, constructed on site with a façade comprising a second skin of untreated larch, the local wood.

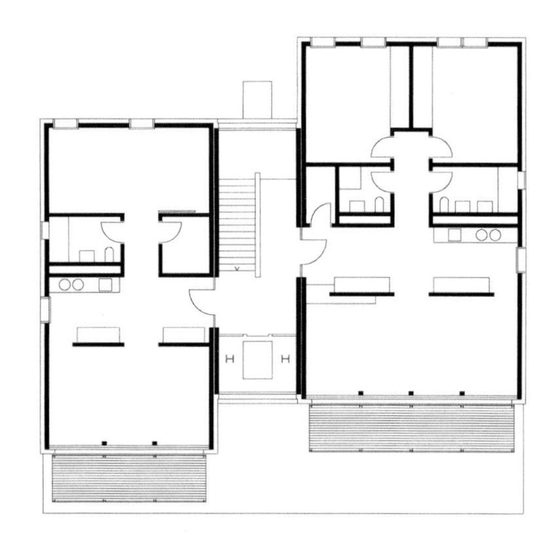

Wood is a particularly important feature of the façades and balconies, overlaying a metal frame, in the treads of the the glass brick and reinforced concrete stairwell, of the apartment interiors – in the parquet flooring and all of the furnishings in the entrance hall and kitchen, spaces that combine treated wood with linoleum.

GALLOWAY RESIDENCE

wood volumes

THE MILLER-HULL PARTNERSHIP
Robert HULL, Petra MICHAELY, Brian COURT
PHOTOS: © Benjamin BENSCHNEIDER

MERCER ISLAND - WASHINGTON - USA

This residence graces a steeped wooded hillside on the shore of Lake Washington. The house is accessed by means of a bridge connected to the uppermost level, and enjoys panoramic views of the lake and its shoreline. A central staircase of wood, steel, and concrete descends cleanly through the house towards the living areas, which are situated on the lowest level and open out to the lake shore.

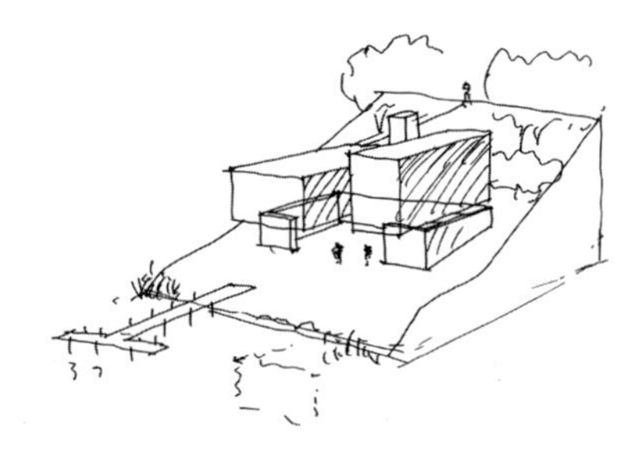

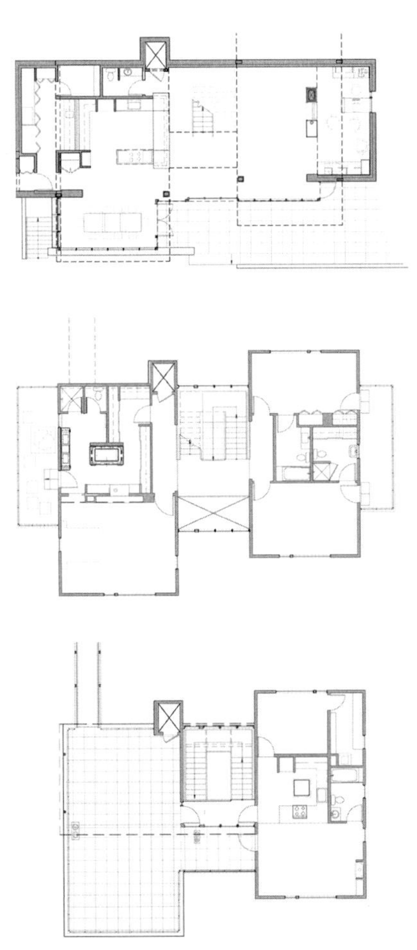

Two large timber-clad boxes rest on a continuous wall of exposed concrete. They contain the most private areas of the design program, extending over the second and third levels. A free-standing house for guests is located on the same level as the uppermost level of the house. The volumes are clad with slats of yellow beech, which also covers the ceiling and walls of the lowest level, lending the structure consistency.

The materials combined in this building are timber cladding, steel, and exposed concrete, always defining the volumes with clarity and taking advantage of views over the forest and the lake. The windows are in matt aluminum and interiors feature apple and bamboo. The house is ventilated naturally, with the central staircase functioning like a courtyard.

CASA EN CHAMARTÍN

second skin

NIETO SOBEJANO ARQUITECTOS, S.L.
COLLABORATORS: Carlos BALLESTEROS, Mauro HERRERO, Juan Carlos REDONDO
PHOTOS: © Luis ASÍN, Hisao SUZUKI

MADRID – SPAIN

The desire to build a new home in a 1950s residential area came into conflict with an ordinance protecting the original building from demolition. The architects significantly transformed the building by means of two design solutions: a new skin wrapped around the volume of the old house, and a central core that penetrated the interior spaces from top to bottom.

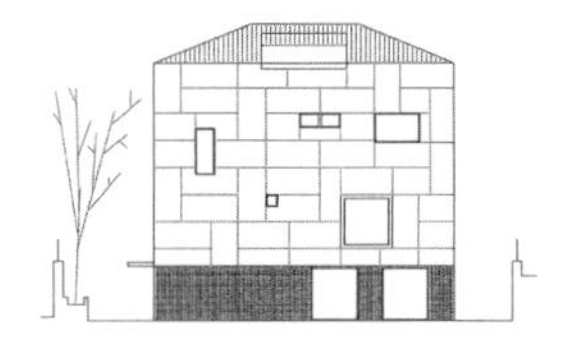
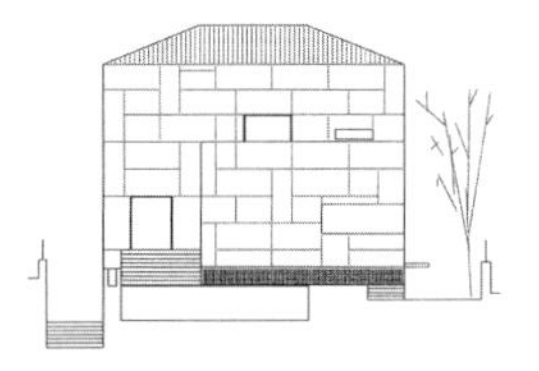
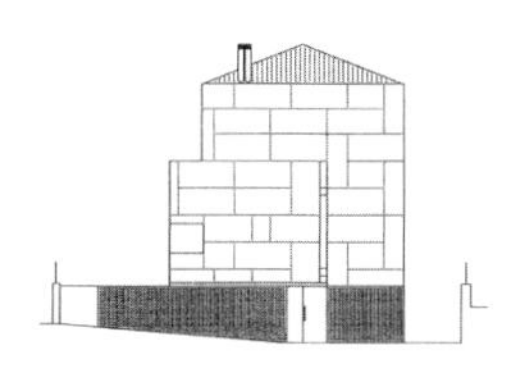
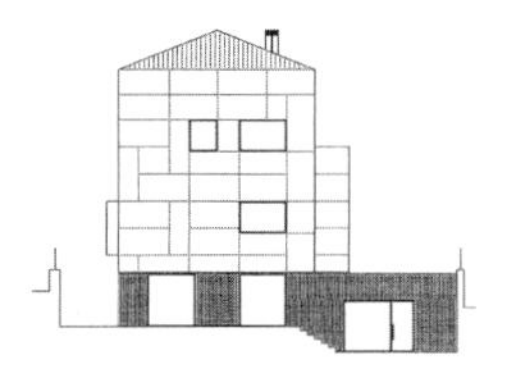

The skin is made of high-density engineered wood boards. This skin is separated from the existing walls by means of an air chamber, and attached to them with aluminum spacers, allowing it to function as a ventilated façade. The lowest level is clad with corrugated aluminum and the roof with copper.

Many of the elements serving the main spaces in the interior of the building are concentrated in a single core unit, the backbone of the house: closets, bookshelves, electrical appliances, service installation, and office conduits, releasing space on all of the floors. Wrapping and centralizing are simple solutions to the problem... perhaps even your lifestyle.

06

sustainable wood

Wood is an ecological, recyclable, and sustainable resource. Here are four projects that include sustainability as an integral design factor. Whether in Brazil or in Central Europe, there is growing evidence of respect for the environment, energy saving, and the awareness that the planet we live on is finite.

CASA BANDEIRA DE MELLO

sustainable habitation

MAURO MUNHOZ ARQUITETURA
COLLABORATORS: Andrea FELTRIN, Eduardo LOPES, Viviane ALVES, Fabiana TANURI
PHOTOS: © Nelson KON

ITÚ – BRAZIL

This house was designed as a weekend retreat for a young couple with two small children and is located 25 miles from their main residence in São Paulo, Brazil. However, it is also used for receiving visitors, holding parties, and family celebrations, with a view to becoming the family's main residence in the future. Consequently, the house is laid out in two wings, one for the family and one for guests.

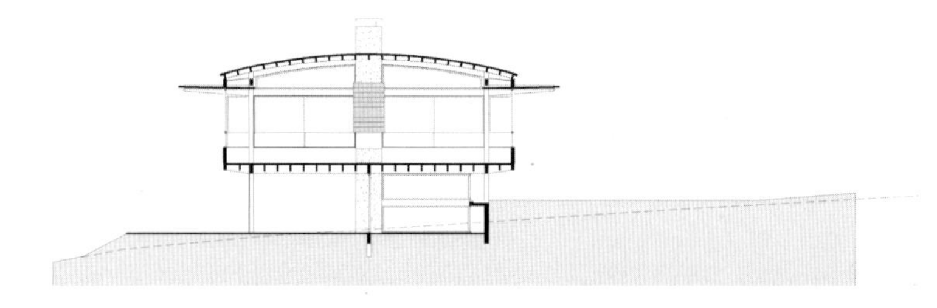

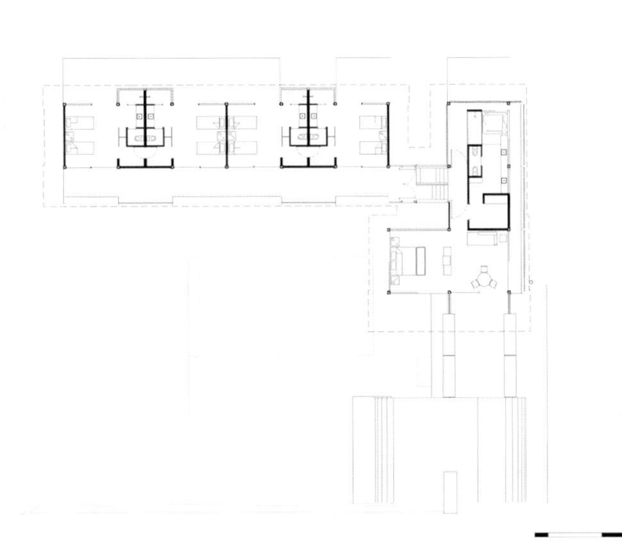

The original topography played a major role in the project. The house is largely transparent from the street, but it respects the occupants' privacy and creates an intimate atmosphere. The home features open spaces resembling porches – the living room, the entrances to the rooms – and makes the most of its privileged tropical location. The sun shines through these intermediate spaces, bathing the rest of the rooms with light.

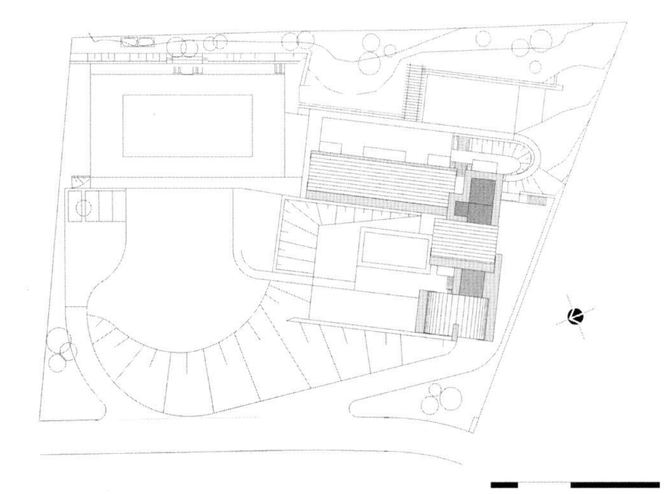

The frame of the house is made entirely of cumaru wood from forests certified by environmental agencies. It was machined and finished in a workshop before being assembled quickly on the site. The design solutions featured in this project, such as passive ventilation, natural lighting, and the use of the sun and the shade provided by the porches and brise-soleil make the house more energy efficient.

STEINWENDTNER HAUS

inner sanctum

HERTL.ARCHITEKTEN
Gernot HELTL
COLLABORATORS: Michael SCHRÖCKENFUCHS
PHOTOS: © Paul OTT

STEYR-MÜNCHHOLZ – AUSTRIA

Designed as a low cost, single family home, this structure was built completely of timber. Only the two structures in the garden, the bicycle shed, and the gardener's tool shed, were made of steel. To make up for the small surface area of the house, the connections to all of the living areas on the lower level are integrated to achieve a space without passageways.

Once inside the house, the general size of the structure is evident, although there is a feel of spaciousness that does not seem possible from the outside. The glazed expanse of the upper story enables sunlight from the south to filter vaporously over the living spaces that are hidden from the street, creating a secluded atmosphere.

On the north side, the living area expands out by means of a terrace, cut away from the body of the structure, towards the neighboring forest. The analogy of the light flooding the house was an intentional and integral part of the project. This light, always filtered by the trees or other rooms, is the same subtle light, full of the nuances that are found in sacred places.

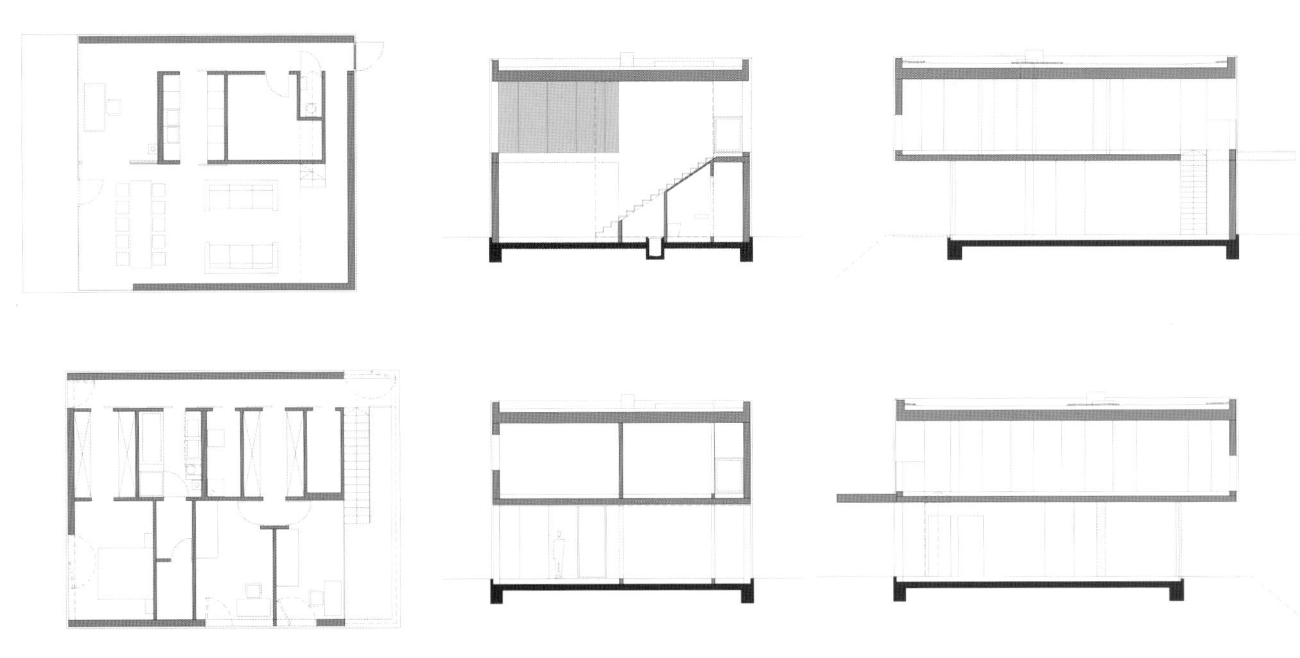

PASSIVEHAUSANLAGE

energy trap

JOHANNES KAUFMANN, OSKAR LEO KAUFMANN

DORNBIRN – AUSTRIA

This row of nine terraced houses and one apartment, all with underground parking, is built entirely of timber, a new development following regulation changes, which in the past stipulated the construction of flame-retardant walls between the different residences. It forms part of a residential area built with much wider blocks; together they create a landscape of replicated façades.

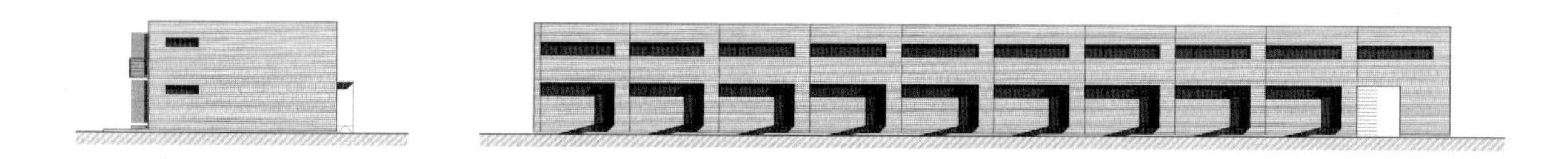

The rational design uses the least amount of space possible, resulting in lower costs. The spaces are small and are open to layout changes depending on the needs of each occupant. The floor plan offers flexibility and the possibility of the redesign of partition walls in the future.

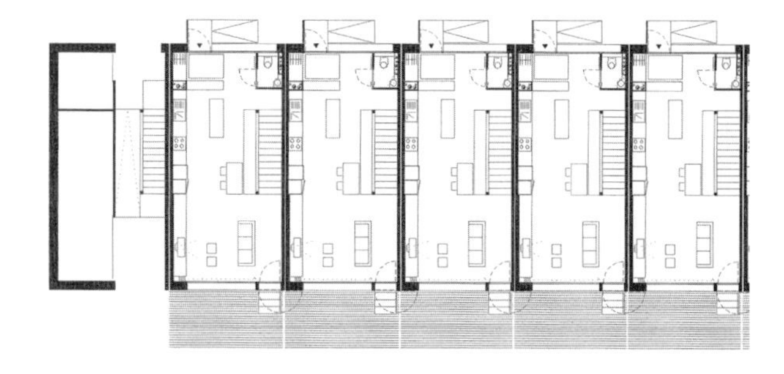

The environmental aspect of the project is very important. The houses are perfectly insulated, with controlled airflow, extremely efficient heating, passive solar capture glazing and skylights, and a ventilated roof, giving the building 60-80% lower energy costs.

WOHNHAUS MIT ATELIER

elegant simplicity

THOMAS MAURER ARCHITEKTURBÜRO

LANGENTHAL - SWITZERLAND

This building contains the family residence and office of its architect. Located in a former industrial area, it comprises sturdy timber modules of an elegant simplicity. The office is in the basement, the living area on the first floor, and the bedrooms upstairs. A porch off the front façade is a unique feature of the building.

Built from fir over a reinforced concrete podium, the edifice has clearly marked north and south-facing frontages. The north side has no windows, whereas the south side is completely open for maximum exposure to the sun. Between them are the service areas containing wet areas and the staircase. Adjustable paneling enables the rooms to be opened out or to be closed off.

Built very quickly, the house consists of a hardwood timber frame, lined with wood panels and clad on the exterior with larch wood slats. This project incorporates energy-saving systems, with passive solar capture, rainwater recycling, and includes the use of locally-sourced wood, free from any polluting treatments.

07

interiors and exteriors

Here is a selection of the many possibilities wood offers for interior design and the decoration of the different rooms of the home.

GREAT (BAMBOO) WALL

KENGO KUMA & ASSOCIATES

PHOTOS: © Satoshi ASAKAWA

The team of architects led by Kengo Kuma built this house near the Great Wall of China for a competition involving promising young Asian architects. The project emphasizes the symbolic significance of bamboo by taking its use to extremes. It was used on outer walls, for partitioning spaces in the house, and to create a warmth that permeates to the exterior.

ROOF HOUSE

TEZUKA ARCHITECTS

PHOTOS: © Katsuhisa KIDA

This residence in Kanagawa, Japan, features a connection between the spacious interior and the exterior that lies beyond the ceiling skylights by means of a ladder in each space. The roof, sloping to match the surrounding landscape, is made comfortable with benches, a table, a kitchen area, and even a shower to become an extension of the house. An outside staircase links the roof directly with the garden.

MICHAELIS HOUSE

HARRY LEVINE ARCHITECTS

PHOTOS: © Patrick BINGHAM-HALL

This house in Sydney, Australia, features a comfortable, spacious, and welcoming living area that is full of subtle beauty. Open to the exterior through the continuous glazed expanse, a dialog is created between the red cedar, the cherry, the Tasmanian oak, and the recycled wood used throughout the house, and the beautiful view of old and wizened trees in the reserve skirting nearby Willoughby Bay and the cliffs of Cammeray.

TREASURE PALACE

EDGE DESIGN INSTITUTE LTD.

PHOTOS: © Janet CHOY, Gary CHANG

This residence in Taipei, Taiwan, features an intelligent space controlled by hidden technology and different layers, where the activites carried out in each space are defined by the objects contained in it. Doing away with the conventional concept of "the living room", the house has "areas", designed for the different daily activities. There is a "green area", where one can relax, at the end of a pebble walkway next to the windows of the house, or an "information area", where one can retreat into boxes made from golden wood, surrounded by books.

MAISON GOULET

SAÏA BARBARESE TOPOUZANOV ARCHITECTES

PHOTOS: © Marc CRAMER

This residence in Quebec, Canada, features a stone slab base from where the timber frame of the house rises, further strengthened by two stone fireplaces that embrace the areas of the house for relaxing and socializing. The wood-paneled interior absorbs and reflects the ambience of the exterior space. The height of the house increases in harmony with the skyward reach of the trees. The house is extended transparently and infinitely by the enchanting vision of the forest, perceiving its trembling and shuddering, and the sound of its whispers and lamentation.

VILLA LÅNGBO

OLAVI KOPONEN ARCHITECT

PHOTOS: © Jussi TIAINEN

This Finnish house was designed not to have barriers to separate its inhabitants from nature. The walls are all but invisible – the glass is only there to keep out the cold. The timber window-frames harmoniously blend in with the tall, straight profile of the trees. Walking through the house is like walking through the forest, while sitting in it is like relaxing beneath the treetops.

CHUN RESIDENCE

CHUN STUDIO, INC.

PHOTOS: © Tim STREET-PORTER

The interior of this home in Santa Monica, California, the external appearance of which is described in more detail in the chapter on exteriors, features calm and peaceful spaces, seeking to reflect the beauty of the landscaped gardens through the use of wood and intentionally understated and warm furnishings.

CASA ACAYABA

MARCOS ACAYABA ARQUITETOS

PHOTOS: © Nelson KON

Set in Brazil's Atlantic forest, this house was designed to be like another tree in the landscape, enabling it to be embedded naturally into its surroundings, and making it possible to become a focal point in the spaces leading to the ocean and to the mountains. Its height, raised to that of the forest, gives unique and privileged respite among the treetops. The interior features abundant wood, turning it into a series of balconies open to the view around it.

CASA BANDEIRA DE MELLO

MAURO MUNHOZ ARQUITETURA

PHOTOS: © Nelson KON

Described in more detail in the exteriors section, this residence in Itú, Brazil, features wood in a wealth of nuances and uses in its unique living areas. These areas are open plan of such spaciousness that the communal areas seem to serve more as a series of connecting terraces than as individual sections.

PINE FOREST CABIN

CUTLER ANDERSON ARCHITECTS

PHOTOS: © Undine PRÖHL

The design of this small cabin in Washington State was meant to fit harmoniously in its setting by minimizing its dimensions. The functional rooms of the house, the bedroom, kitchen, and bathroom are located in the more hidden section of the house, while the double height living area, its façade simply held together with timber posts, increases the interior space of the home by becoming a part of the exterior.

RESIDENTIAL IN ÖSCHINGEN

SCHLUDE ARCHITEKTEN

PHOTOS: © Martin RUDAU

This residential project in Germany shows a firm support for energy saving by making good use of passive solar energy. The design of the simple and compact volumes, clean lines exuding restraint, and quality construction give the interiors a simple but welcoming feel. Wood is extensively used, in this case, as the result of its sustainable and ecological nature, and owing to its origin, performance, and recyclability.

WOHNHAUS STUTTGART

SCHLUDE ARCHITEKTEN

PHOTOS: © Juraj LIPTAK, Reiner BLUNK

This residential project in Germany creates a large multi-purpose space, which can be redefined with moveable paneling and offers transparency and privacy through the light being softened by the slats of the blinds. The timber features filter light from outside, casting shadows as it passes through the treads of the staircase.

CHALET PICTET

CHARLES PICTET ARCHITECTE

PHOTOS: © Francesca GIOVANELLI

This solid timber chalet, built in the Swiss Alps in 1872 and restored and extended at a later date, reflects the slow passage of time in its interior, like the drops of a fine and everlasting rain. The house features a fusion of past and contemporary styles by contrasting the patina of the aged wood with the newly cut timber, highlighting the touch of what has gone before as it blends with the pulse of modern life. The warmth and inviting nature of the wood pervade the ambience of this space contained by sturdy walls.

DUPLEX RUE DE SAVOIE

LITTOW ARCHITECTES

PHOTOS: © Littow ARCHITECTES

This remodeling of a Paris loft restored the original timber frame, cleansing it of all additional construction materials. The ceiling undulates like the waves of the sea, inspiring calm that pervades the space below. The horizontal planes are connected vertically by the timber beams, which also serve their own purpose of dividing the room into two separate spaces.

MEHRFAMILIENHAUS II

VOGT ARCHITEKTEN

PHOTOS: © Dominic BÜTTNER

Closets of varying heights combine the different shades of wood and linoleum in a dark color spectrum, in contrast with the parquet flooring, to give this kitchen a simple and elegant feel, one of such order and cleanliness that it seems too good to cook in.

CHALET PICTET

CHARLES PICTET ARCHITECTE

PHOTOS: © Francesca GIOVANELLI

Enveloped in wood, the kitchen of this restored home is a place where that passage of time has no importance. The warmth of ready to eat dishes and the smells of the food being cooked slowly and purposefully over a low fire appear to hide forever in the solid and polished walls that enclose it, lined in wood full of half closed eyes.

LATORRE RESIDENCE

GARY CUNNINGHAM ARCHITECT

PHOTOS: © Undine PRÖHL

This house in Dallas features a dynamic and contemporary kitchen comprising compact units that give it a somewhat ironic sense of efficiency, as can be seen by the spotlights focusing on each unit. The timber beams in the ceiling are completely integrated into the kitchen space, to the point where they serve as anchoring points for hanging shelves and cooking utensils.

WOOD RESIDENCE

CUTLER ANDERSON ARCHITECTS

PHOTOS: © Undine PRÖHL

This kitchen appears at the end of a flowing space shared with the living and dining areas; a space imbued with character by the heat of the wood stove, the divider between the two areas. Entering this place, we delve into a rich forest of different tree species, of closets that could be a refuge for birds of different types, and beams that seem to wait for a bird to roost after its flight, while time passes lazily by in the midst of warm colors and soft lighting.

MICHAELIS HOUSE

HARRY LEVINE ARCHITECTS

PHOTOS: © Patrick BINGHAM-HALL

In this Australian home, a row of wood closets and shelves frame the landscape that is the kitchen. The glazed expanses that let natural light into the kitchen also support shelves displaying X-ray images of everyday kitchen objects.

MAISON GOULET

SAÏA BARBARESE TOPOUZANOV ARCHITECTES

PHOTOS: © Marc CRAMER

This interior design project focuses its attention on the continuous presence of wood paneling on the walls. The kitchen compartment hides behind an open stone partition wall, and also behind an open countertop, separating its occupants from the expanse of glass and wood used to build the entire house, which turns it into a pleasant walk through the natural landscape.

ISLAND HOUSE

ARKITEKTSTUDIO WIDJEDAL RACKI BERGERHOFF

PHOTOS: © Åke ERIKSON LINDMAN

Gazing out over the ocean between oak trees and rocky outcrops there is a space for socializing, of which the kitchen forms an essential part. Encased by and blending discreetly into the wood that covers the entire house, the kitchen is like a lady silently displaying her elegance in the way it is almost merged into the wall that it is built against.

STUDIO 3773

DRY DESIGN

PHOTOS: © Undine PRÖHL

In small homes, such as this one in Los Angeles, innovation is often necessary to make the most of the available space. Here, the design solution for the bedroom was a sleeping loft over the living area, esconced in the beams of the frame – a charming feature that uses cleverly-worked timber to enhance the space.

JOHN LE CARRÉ
THE CONSTANT GARDENER
DIAMOND
MATTHEW HART

DUPLEX RUE DE SAVOIE

LITTOW ARCHITECTES

PHOTOS: © Pekka LITTOW

The remodeling of this Paris loft enabled the attic to be turned into a bedroom. The timber frame, completely stripped and restored, brings warmth to what was once an empty space. This sense of the temporary, like a migrating bird, is enhanced by the bed, which has been placed – almost monastically – directly on the floor, illuminated by skylights.

Chalet PICTET

CHARLES PICTET ARCHITECTE

PHOTOS: © Francesca GIOVANELLI

The three bedrooms in this chalet exemplify three different ways of using wood. The sinuous wood blends function with adornment in the wall, forming a mosaic of circumferences with the grain and hues found in the material. Aged over time, timber brings a simple, rustic, and understated feel to the home. The wood has been adapted to modern times to create an air of simplicity, while reducing the interior to one unique space.

TREASURE PALACE

EDGE DESIGN INSTITUTE LTD.

PHOTOS: © Janet CHOY, Gary CHANG

This bedroom is an updated version of Oriental-style interior design and features strict directionality. The room is laid out as a continuous space where the furnishings rather than the walls bring definition. Taken to the extreme, the wall separating the bedroom from the passageway becomes a simple partition; its slats allow light to enter, creating a beautiful image.

MAISON GOULET

SAÏA BARBARESE TOPOUZANOV ARCHITECTES

PHOTOS: © Marc CRAMER

This bedroom is a simple geometric space free of details. The bed is positioned in the center of a wood-lined box, with the light entering through small openings from the immensity of the landscape that lies beyond to wash the space with serenity. The bed is the only object alluding to the function of the room, while it creates a silent dialogue with the autumnal hues of its timber surrounds.

FLOODED HOUSE

GAD ARCHITECTURE

PHOTOS: © Ali BEKMAN, Salih KUCUKTUNA

Constructing a present in this century without rejecting heritage; giving continuation to history without reproducing the past; designing without imitating, using the salvaged parts of an old body. Vestiges of the cultures that form the rich patchwork that is the city of Istanbul today pervade these bedrooms in two houses from far-off and very different times. A bed rests under the skeletal frame of an Ottoman house; another is positioned next to a wall that has witnessed the sleeping and waking of three generations.

HOUSE IN FINLAND

WOOD FOCUS FINLAND

PHOTOS: © Esko JÅMSÅ, Mikko AUERNIITTY

In this Finnish home, timber slats have been used to line the walls and ceiling, painted blue to give a maritime ambience. Contrasting horizontal and vertical lines in the surfaces and furnishings are a feature of this space devoted to repose.

TREASURE PALACE

EDGE DESIGN INSTITUTE LTD.

PHOTOS : © Janet CHOY, Gary CHANG

One of the hidden treasures of this home is the bathroom. In one of the spaces in this palace, an immense field of comfort and pleasure where the richness of the wood is bright with a golden sheen, one can find simple fixtures outlined in the grooved surfaces characterizing the bathroom.

SUITCASE HOUSE HOTEL

EDGE DESIGN INSTITUTE LTD.

PHOTOS : © Asakawa SATOSHI, Howard CHANG, Gary CHANG

The bathroom is hinted at, half-buried in the long continuous space of this changing house located next to the Great Wall of China. It appears and disappears beneath a floor of wooden hatches that open and close, to provide the necessary privacy or to form other spaces.

CASA BANDEIRA DE MELLO

MAURO MUNHOZ ARQUITETURA

PHOTOS : © Nelson KON

The topographical nature of the site where the Brazilian Bandeira de Mello home is located allows floor-to-ceiling windows to be a feature of the bathroom. Its warmth and simplicity contrast with the somber and introverted feel of its counterpart in Island House, which discreetly and elegantly shares its intimacy only with the Swedish sky.

ISLAND HOUSE

ARKITEKTSTUDIO WIDJEDAL RACKI BERGERHOFF

PHOTOS: © Åke ERIKSON LINDMAN

HOUSE WITH THREE ERAS

GAD ARCHITECTURE

PHOTOS: © Ali BECKMAN, Salih KUCUKTUNA

Stone, earth, wood, brick, and concrete impregnate this bathroom with the colors of fall, and grooved surfaces, to create an extensive space with simple fixtures seemingly suspended in midair. The volume that encases this space is a patchwork of geometric layers created by different colors and materials. Unassumingly, wood is used specifically and coordinates well with the other materials.

HOUSE R

FEYFERLIK/FRITZER

PHOTOS: © Paul OTT

Bathrooms are often dark and claustrophobic spaces. In this finished project in Austria, this limitation has been overcome by replacing the walls enclosing this space with shelving, which enables light from the outside to penetrate completely. The strategic placement of books can help to do away with the need for privacy or lighting, to the user's taste. White ceramic bathroom fittings were chosen to contrast with the wood.

HOUSE IN FINLAND

WOOD FOCUS FINLAND

PHOTOS: © Esko JÅMSÅ, Mikko AUERNIITTY

On entering this space, attention is immediately drawn to the wooden sauna, which invites us to enter the warmth and tranquility inside through doors that prolong the harmonious feel of the bathroom. Once again, the color of the wood offsets the white bathroom fixtures. The windows give a feeling of spaciousness and comfort.

CHALET PICTET

CHARLES PICTET ARCHITECTE

PHOTOS: © Francesca GIOVANELLI

Here is another look at this remodeled chalet in the Swiss Alps. The warm and gentle color and the smooth texture of the planed wood previously existing in the structure contrast with the grayish, rough-hewn stone, enhancing the warmth of the bathroom.

PORCH HOUSE

TEZUKA ARCHITECTS

PHOTOS: © Katsuhisa KIDA

The outside is the inside, and the inside is the outside. The whole house is a giant porch, bringing privacy through the embrace of the surroundings. The breeze is the oxygen its inhabitants breathe, the light of the sun is the lighting that paints the chiaroscuro of its corners. This is life in the open air without heat or cold, languishing beneath the stars and bathing under the rain.

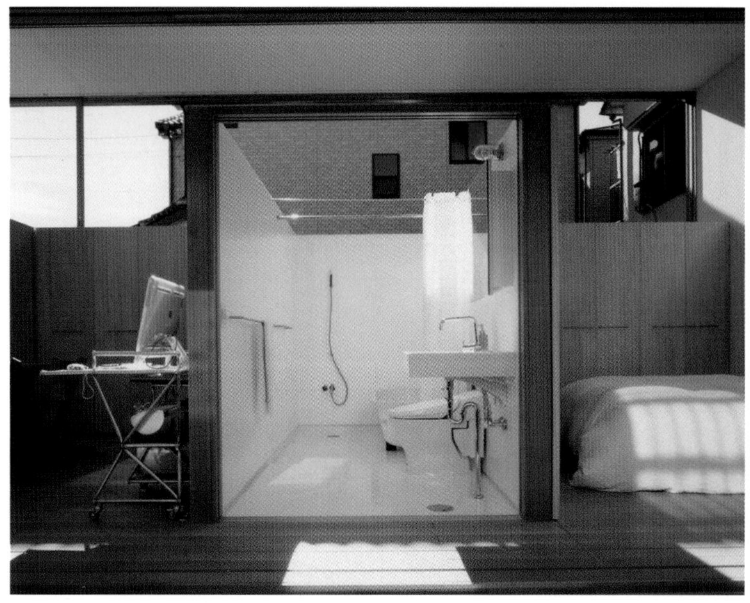

AVALON HOUSE

CONNOR + SOLOMON ARCHITECTS

PHOTOS: © Peter SCOTT

This house in Australia is a clear example of how a small terrace, which would be an afterthought in other projects, is effectively optimized. A narrow terrace is given an important role by means of the transparent façade, amplifying the interior and the terrace. A large area running the width of the house, raised slightly above the ground, creates cozy spaces that are close to nature and close to home.

HAUS NENNING

CUKROWICZ-NACHBAUR ARCHITEKTEN

PHOTOS: © Hanspeter SCHIESS

This residence in Austria, described in the chapter on exteriors, features outside space contained in a box with see-through doors and timber-framed windows that allow the light to peek through, hinting at the day. Doors and windows open like those of a storefront, generously displaying what cannot be bought, but only given. The wood tinges the areas in between with a warm light.

HAUS HEIN

CUKROWICZ-NACHBAUR ARCHITEKTEN

PHOTOS: © Albrecht SCHNABEL

The traditional-style house in Fraxern, Austria, has a contemporary twist to it by way of a platform that takes the occupants closer to the view. Intersecting the vertical line of the house, this terrace features a covered area and a balcony holding up the sky. The floor is a continuous line that leads the eye to the horizon, mountain scenery that grows ever distant but never quite disappearing.

FLOODED HOUSE

GAD ARCHITECTURE

PHOTOS: © Ali BEKMAN, Salih KUCUKTUNA

On the banks of the Bosphorus between Europe and Asia, reminiscent of the splendour of the Ottoman period and continuing the story to today, the skeletal frame of this house reappears incarnated in a living specter standing between two seas, two continents, and many centuries that have been, and that have yet to pass. Its porch and terrace harbor a calm space overlooking the slow movement of the Bosphorus and its striking proximity, seeming to merge with the water of the sunken swimming pool that penetrates almost to the interior of the house.

GREAT (BAMBOO) WALL

KENGO KUMA & ASSOCIATES

PHOTOS: © Satoshi ASAKAWA

The subtlety of bamboo, symbolizing the cultural exchange between China and Japan, forms a window in the impenetrable mass of the Great Wall. The intertwined bamboo canes form the exterior space of the house, contained in the interior like brushstrokes on the whiteness of the paper, without obstacles or depth.

MATTHEWS RESIDENCE

JMA ARCHITECTS

PHOTOS: © John MAINWARING

This residence in Bardon, Australia, features two projecting wooden terraces that grow with the height of the house. They are beautifully supported by two pillars, the sloping of which symbolizes the reaching out from the interior to the surrounding forest, returning to their origin.

08

dreaming

At the end of the road, we return to our starting point. Wood on wood. Tree houses make our childhood dream of independence come true; they are our hideaway, our contact with nature. Here are three examples, one clearly a place to dream, one for rational recreation, and a third that is almost a real home.

PAULMEAD TREEHOUSE

new life

TOM MILLAR, TOM HOWARD, JAMES FOOTTIT
PHOTOS: Melissa MOORE

BISLEY – UNITED KINGDOM

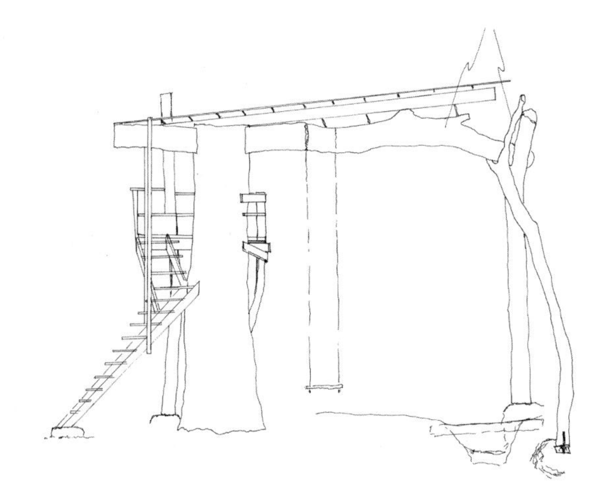

Giving use to an old oak tree, nothing now but a skeleton, and breathing life into its remains without tearing it from the ground are behind the design of this shelter mainly designed for the children and grandmother of the family. The trunk serves as the central pillar of this refuge for long lost stories and feelings.

The stairway leads to a little hideaway, almost completely open except for the rail and a roof that is little more than a sunshade. It is a dream factory. While it has no walls, the roof is wide enough to protect the interior from the rain; an interior with sufficient space to host any kind of unannounced visit.

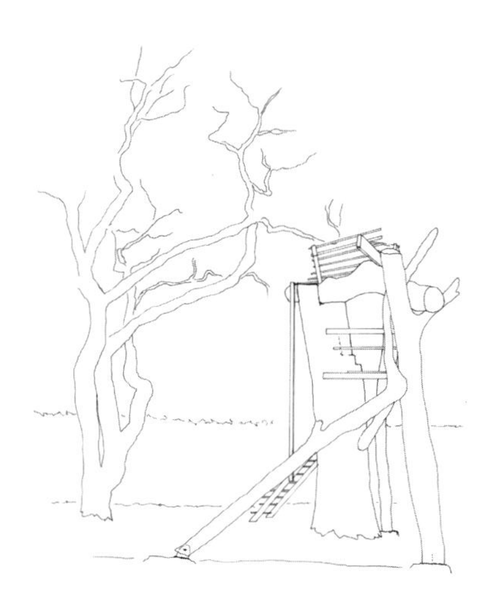

In time, the new elm tree planted next to it will form branches to reach and surround the tree house. Shoots and flowers will bestow it with color, scent, and a new life. Birds have already come to nest in it. The thick plant life around the tree creates the necessary image of mystery and separation of the garden… A swing hangs from a branch.

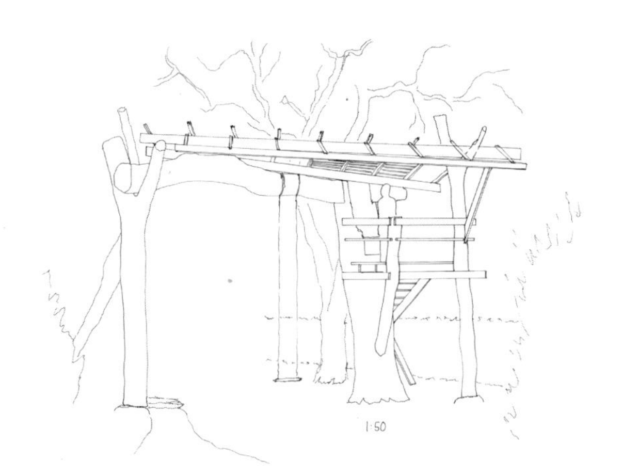
1:50

LEE TREEHOUSE

negotiating a coexistence

JOSEPH LIM
COLLABORATORS: C.I. YEO, Lai WAI HENG
PHOTOS: © Ismurnee KHAYON

GALLOP PARK – SINGAPORE

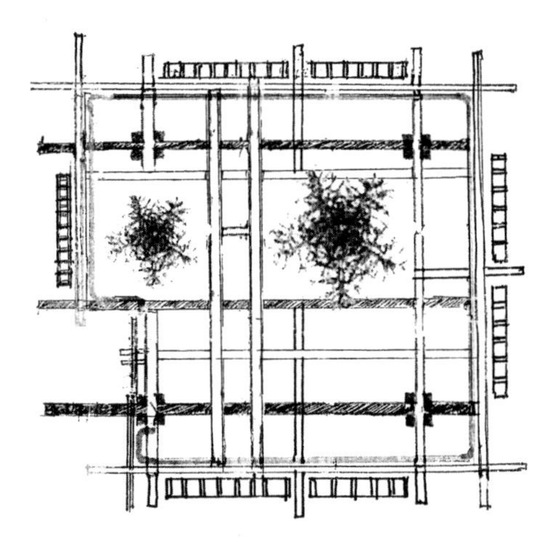

The aim of this tree house project was to negotiate a wood and metal structure to coexist with two existing trees. The effort involved designing a sound structure that respected the ground rules laid down by the trees. The first stage of development was to create a detailed scale drawing of the trees.

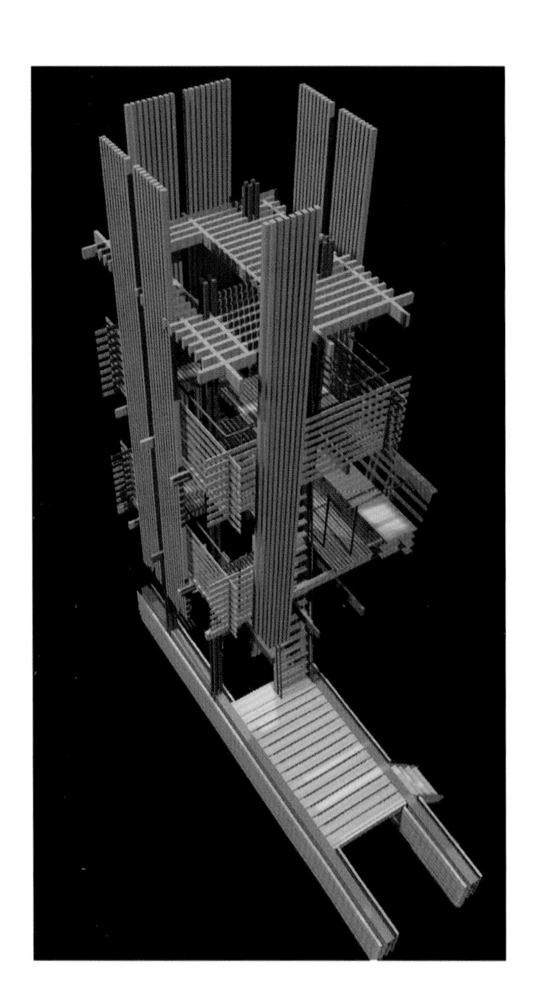

It was essential to obtain a strict and precise definition of the tree in order to build the tree house, involving identification of the fundamental points of support. The structure had to sprout from within the old body, giving it new life. A juxtaposition of wooden beams and iron hugs the entire tree to create a delightful structure that communicates between nature and the present.

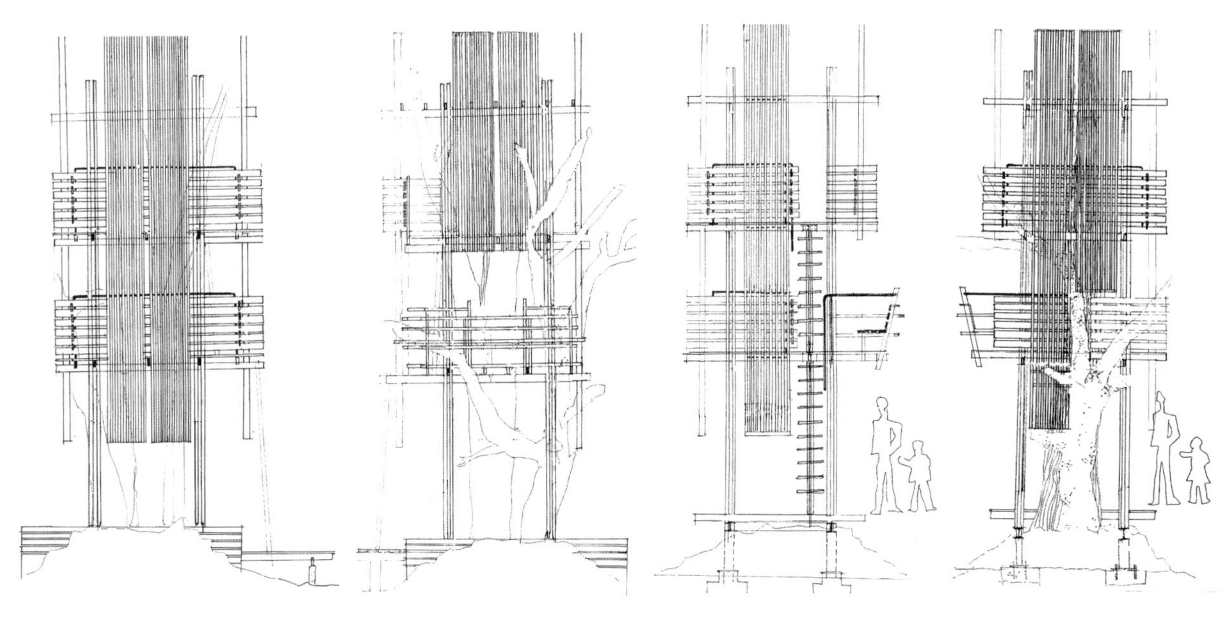

The interior is a maze of secret niches, filled with the excitement of children seeking a place in nature that accommodates every little aspect of their imagination. To make this happen, the architect attempted to reconnect with childhood, turning the tree house into a solid launching pad for a magic carpet to take you anywhere you want to go. You are never too old to climb into the trees with the children who share your dreams.

JOHN HARDY HOUSE

double-decker

CHEONG YEW KUAN
PHOTOS: © Rio Helmi

SAYAN – BALI

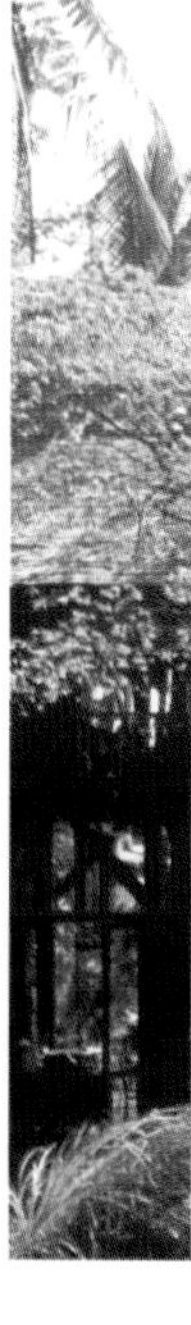

A union with nature and becoming an essential part of her wholeness is one of the elemental desires of man, which hides and reveals itself throughout a person's lifetime. A bird's eye view of the ground, watching the world go by from a cloud, bearing an unseen witness to the events of the world are the ideas that are nursed inside this house nested in the treetops among the branches.

Like a hovering mirage, its horizontal structure cuts across the vertical line of the trees. Struts of treated wood brace its frame, made of cut logs and of trees that are still growing. A subtle roof gives this perfectly habitable house a wild appearance in a patchwork of rough and smooth layers.

With its prominent stilt design, this tree house features an open lower deck covered by a porch that houses the more communal areas, and a upper level that is actually built over the porch, with the more private spaces. The subtle combinations of wood tones, and fabrics and furnishings in warm colors and tans give an immediate sense of comfort.

CATALOG OF WOODS

Information is provided here on the different types of wood, detailing their appearance, nomenclature, physical properties, and applications.

GLOSSARY OF TERMS

· **Xylem (also sapwood):** the area of the tree containing living cells, which transport or store nutrients; it is more porous and clearer in color than the duramen, rotting is unavoidable.
· **Heartwood:** mature wood that is the structural part of the tree; it has no living cells.

Microscopic characteristics

· **Fiber:** a collection of cells that control the direction of tree growth. Straight fibers – parallel to the tree trunk – make the wood easier to work with. Intertwined, wavy, twisting, or crooked fibers are more difficult to work, but have more decorative uses.

Physical characteristics

· **Density:** relationship between mass and volume at 12% humidity, shown in lbs/cu ft. Conifers range from 25 (very light) to 44 (very heavy). Deciduous, from 31 to 60.
· **Hardness:** indicates the ease with which one material will penetrate another, and illustrates the resistance of the wood to abrasion or chipping.
· **Hygroscopicity:** properties of wood for moisture exchange with the atmosphere.
· **Dimensional stability:** decrease in the dimensions of a piece of wood owing to humidity loss.
· **Unitary dimensional stability volume coefficient:** the change in volume experienced by a piece of wood following a 1% change in humidity. According to the coefficient, the wood can be:

Inflexible (0.15-0.40): for use in furniture making.
Flexible (0.35-0.49): ideal for use in joinery.
Quite flexible (0.40-0.55): for use in construction.
Flexible (0.55-0.75): for construction in radial quartering.
Very flexible (0.75-1): for use in conditions of constant humidity.

Mechanical characteristics

· **Split or crack:** indicates the resistance to fixing or screwing or in certain assemblies.

Chemical characteristics

· **Workability:** woods containing more resins, fats, or wax can cause problems in shaping and finishing. Mineral deposits – mineral salts from the ground precipitate to form crystals – again make workability difficult.
· **Natural durability:** the resistance of the wood to attack from destructive agents. This is important when the wood is exposed to changes in humidity higher than 18%. Agents range from fungus, xylophagous insects (coleoptera and termites) to xylophagous marine organisms. Durability refers to the heartwood, since the xylem is always susceptible to attack, and ranges from durable to soft/non-durable.
· **Impregnability:** the ease with which liquids can penetrate the wood. A soft but impregnable wood can resist attacks from xilophagous organisms.

1. Pregnable: easy to impregnate under conditions of pressure.
2. Fairly pregnable: two or three hours of exposure for medium penetration.
3. Fairly impregnable: after three or four hours, only superficial impregnation.
4. Impregnable: practically impregnable.

ABIES alba Mill. (SILVER FIR) Sapin argenté (f) Avezzo (i) Tanne (d) Abeto (e)

European conifer in good supply.

Physical properties: density of 25 lbs/cu ft, quite flexible, soft (1.4), fractures easily, contains very few resins, good humidity resistance and comprised of straight fibers. Deciduous from Europe and Asia, in good supply.

Technical properties: easily glued, nailed, and soft screwed. Allows for good finishing, but does not absorb different colors and intensities uniformly.

Natural durability: low durability (fungi) and sensitive (insects).

Impregnability: fairly pregnable or fairly impregnable heartwood, fairly pregnable xylem.

Applications: interior joinery, framework, decorative veneer, furnishing, and glued-laminate timber.

ACER pseudoplatanus L. (SICOMORE) Erable (f) Acero montano (i) Bergahorn (d) Arce (e)

Evergreen from Europe and Asia, in good supply.

Physical properties: density of 38-43 lbs/cu ft, quite flexible, fairly durable (4.7), very resistant to abrasion. Straight or intertwined fibers. Prone to bending with steam.

Technical properties: easily glued, nailed, and screwed. Good finish.

Natural durability: non-durable (fungi) and sensitive (insects).

Impregnability: pregnable.

Applications: interior joinery and woodwork (flooring), decorative veneer, furnishing, and cabinetmaking.

ANCOUMEA klaineana Pierre (OKUME)

Deciduous from Central Africa, in good supply.

Physical properties: density of 27-28 lbs/cu ft, inflexible, soft (1.5). Straight or intertwined fibers.

Technical properties: easily glued, nailed, and screwed. The finish requires a preliminary sealer treatment.

Natural durability: low durability (fungi) and sensitive (insects).

Impregnability: fairly impregnable heartwood.

Applications: plywood boards, interior joinery, decorative veneer, furnishing, and cabinetmaking.

BETULA alba L. (BIRCH) **Bouleau blanc (f) Betula blanca (i) Weissbirke (d) Abedul (e)**

European conifer in good supply.

Physical properties: density of 25 lbs/cu ft, quite flexible, soft (1.4), fractures easily, contains very few resins, good humidity resistance, and comprised of straight fibers. Deciduous from Europe and Asia, in good supply.

Technical properties: easily glued, nailed, and soft screwed. Allows for good finishing, but does not absorb different colors and intensities uniformly.

Natural durability: low durability (fungi) and sensitive (insects).

Impregnability: fairly pregnable or fairly impregnable heartwood, fairly pregnable xylem.

Applications: interior joinery, framework, decorative veneer, furnishing, and glued-laminate timber.

BETULA alleghaniensis Britt. (YELLOW BIRCH) **Bouleau jaune (f) Betula gialla (i) Gelbbirke (d) Abedul amarillo (e)**

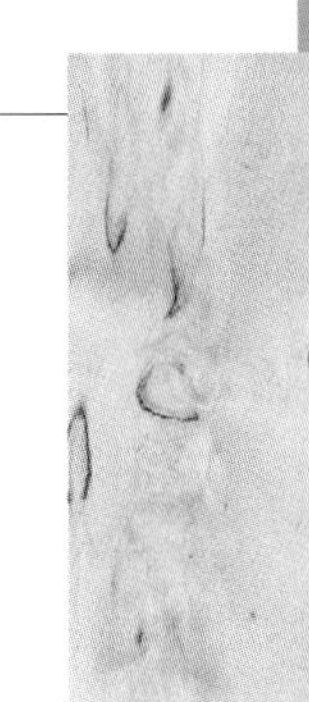

Deciduous from Canada and the United States, in limited supply.

Physical properties: density of 34-44 lbs/cu ft, quite flexible, fairly durable (more so than birch). Straight fibers. Easy to work, good resistance to flexing, compression, and impact.

Technical properties: easily glued, nailed, and screwed. Pre-drilling. Good finish.

Natural durability: non-durable (fungi) and sensitive (insects).

Impregnability: pregnable and fairly impregnable.

Applications: plywood boards, high quality joinery, decorative veneer, furnishing, flooring.

CASTANEA sativa Mill. (SWEET CHESNUT) **Châtaigner (f) Castagno (i) Edelkastanie (d) Castaño (e)**

Deciduous from Europe (Mediterranean basin) and Asia, in fair supply.

Physical properties: density of 34-40 lbs/cu ft, inflexible, soft (2.1). slightly wavy fibers, accelerates metal corrosion in humid conditions and stains blue.

Technical properties: durable (insects), fairly resistant to termites.

Natural durability: durable (fungi), fairly durable to termites.

Impregnability: pregnable heartwood, fairly pregnable xylem.

Applications: joinery and woodwork (doors, windows, and flooring), cabinetmaking.

CHAMAECYPARIS nootkatensis Spach (YELLOW CEDAR) Ciprés du nutka (f) Cipresso americano (i) Nutka Scheinzypresse (d) Cedro (e)

Deciduous from North America, in fair supply.

Physical properties: density of 27-33 lbs/cu ft, inflexible, fairly durable. Straight fibers. Acid resistant.

Technical properties: easily glued, nailed, and screwed, good finish.

Natural durability: fairly durable (fungi) and sensitive (insects), resistant to xylophageous marine organisms.

Impregnability: fairly pregnable heartwood, pregnable xylem.

Applications: interior and exterior woodwork, decorative veneer, furnishing, and cabinetmaking.

CHLOROPHORA excelsa Benth (IROKO)

Deciduous from Africa, in good supply.

Physical properties: density of 39-42 lbs/cu ft, quite flexible, fairly durable (4), straight fiber.

Technical properties: easily glued, nailed, and screwed. Finish: requires preliminary sealer.

Natural durability: very durable (fungi), resistant to termites, and sensitive to xilophagous marine organisms.

Impregnability: impregnable heartwood, pregnable xylem.

Applications: structural plywood boards, interior and exterior woodwork, framework, decorative veneer, furnishings and cabinetmaking, glued laminated timber. Serves as a substitute for teak.

ENTANDROPHRAGMA cylindricum Sprague. (SAPELE)

Deciduous from Africa, in fair supply .

Physical properties: density of 40-44 lbs/cu ft, quite flexible, fairly durable (3.6-4.2). Intertwined fibers, can exude resin, and presents a grooved profile.

Technical properties: easily glued, nailed, and screwed, good finish.

Natural durability: fairly durable (fungi), resistant (insects), sensitive to xylophageous marine organisms.

Impregnability: fairly pregnable heartwood, pregnable xylem.

Applications: interior and exterior woodwork, decorative veneer, furnishing, and cabinetmaking.

FAGUS sylvatica L. (COMMON BEECH) **Hêtre commun (f) Faggio (i) Gemeine Buche (d) Haya (e)**

Deciduous from central Western Europe, in good supply.

Physical properties: density of 43-47 lbs/cu ft, quite flexible, fairly durable (4), straight fibers.

Technical properties: easily glued, nailed, and screwed. Finish: preliminary sealing required.

Natural durability: non-durable (fungi) and sensitive (insects).

Impregnability: pregnable.

Applications: interior woodwork (flooring), decorative veneer, furnishings, and (bentwood) cabinetmaking.

FRAXINUS excelsior L. (ASH) **Frêne (f) Frassino (i) Esche (d) Fresno (e)**

Deciduous from Northern Africa, East Asia, and Europe, in good supply.

Physical properties: density of 42-47 lbs/cu ft, flexible, fairly durable (4-5.3), straight fibers, bends easily.

Technical properties: easily glued, nailed, and screwed with pre-drilling, good finish.

Natural durability: non-durable (fungi) and sensitive (insects).

Impregnability: pregnable.

Applications: decorative veneer, flooring.

HYMENEA spp. (JATOBÁ)

Deciduous from South America, in good supply but with limited production.

Physical properties: density of 40-44 lbs/cu ft, quite flexible, fairly durable (3.6-4.2). Intertwined fibers, can exude resin, and presents a grooved profile.

Technical properties: easily glued, nailed, and screwed, good finish.

Natural durability: fairly durable (fungi), resistant (insects), sensitive to xylophageous marine organisms.

Impregnability: fairly pregnable heartwood, pregnable xylem.

Applications: plywood, interior and exterior woodwork, decorative veneer, furnishing, and cabinetmaking.

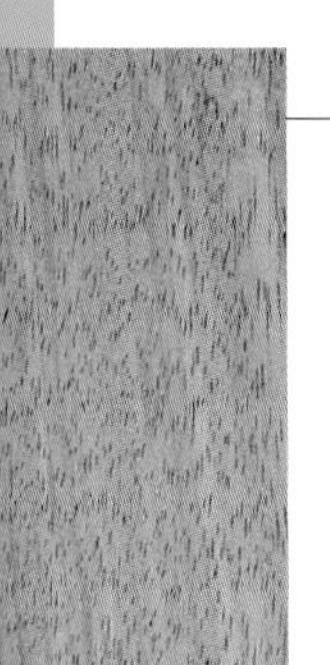

INTSIA bijuga O. Ktze. L. (MERBAU)

Evergreen from Southeast Asia and Australaia, in limited supply.

Physical properties: density of 46-52 lbs/cu ft, quite flexible, hard (6.4), intertwined fibers, oily to the touch, contains rubber, silica and sulfurous deposits stains on contact with metals.

Technical properties: easily glued, nailed, and screwed with pre-drilling. Finish: requires preliminary surface preparation.

Natural durability: very durable (fungi) and fairly durable (insects).

Impregnability: pregnable.

Applications: exterior (windows) and interior (flooring, stairs) woodwork, framework, decorative veneer, furnishing, and cabinetmaking. Similar to afzelia.

JUGLANS regia-nigra L. (WALNUT TREE) **Noyer (f) Noce (i) Walnuss (d) nogal (e)**

Deciduous from North America, Europe, Asia, and Northern Africa, in good supply.

Physical properties: density of 34-42 lbs/cu ft, quite flexible, fairly durable (3.2-3.6), straight fibers.

Technical properties: easily glued, nailed, and screwed, good finish.

Natural durability: fairly durable (fungi) and sensitive (insects).

Impregnability: fairly pregnable heartwood, pregnable xylem.

Applications: interior woodwork (paneling, flooring), decorative veneer, furnishing, and cabinetmaking.

KHAYA ivorensis A. Chev. (MAHOGANY) **Acajau (f) Mogano (i) Mahagoni (d) Caoba (e)**

Deciduous from Western Africa, in good supply.

Physical properties: density of 30-33 lbs/cu ft, quite flexible, soft (1.9), straight fibers.

Technical properties: easily glued, nailed, and screwed. Finish: preliminary sealing required.

Natural durability: fairly durable (fungi) and sensitive (insects).

Impregnability: impregnable heartwood, pregnable xylem.

Applications: plywood boards, interior and exterior woodwork, decorative veneer, furnishing, and cabinetmaking. Can be used to substitute Sipo and Sapele.

LARIX dedicua Mill. (LARCH) **Melèze (f) Larice (i) Lärche (d) Alerce (e)**

Conifer from Central Europe and North America, in good supply.

Physical properties: density of 29-40 lbs/cu ft, flexible, fairly durable (2.2-3.2), straight fibers, resin canals.

Technical properties: easily glued, nailed, and for fine screws. Finish: resinous.

Natural durability: low durability (fungi) and sensitive (insects).

Impregnability: impregnable heartwood, fairly pregnable xylem.

Applications: plywood boards, cladding, interior woodwork (flooring), decorative veneer.

MILLETIA laurentil De Wild (WENGÉ)

Deciduous from Africa, in fair supply.

Physical properties: density of 44-56 lbs/cu ft, flexible, hard (9), straight fibers, resistant, and resinous.

Technical properties: lightly glued, nailed, and screwed with pre-drilling holes, finished with wax-based products.

Natural durability: durable (fungi) and sensitive (insects).

Impregnability: impregnable heartwood.

Applications: interior joinery and exterior woodwork (flooring), decorative veneer, furnishing, and cabinetmaking.

OLEA europaea L. (OLIVE TREE) **Olivier (f) Ulivo (i) Olivenbaum (d) Olivo (e)**

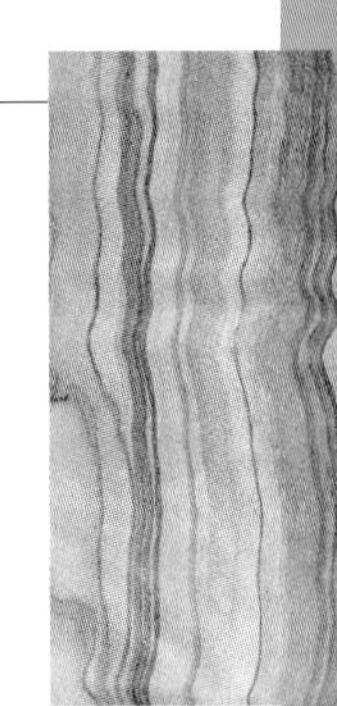

Deciduous from the European Mediterranean basin, in fair supply, limited plantations.

Physical properties: density of 53-70 lbs/cu ft, flexible, hard, uneven fibers, oily to the touch.

Technical properties: not easily glued (oils), nailed, and screwed with pre-drilling, good finish.

Natural durability: durable.

Impregnability: fairly pregnable heartwood, pregnable xylem.

Applications: interior woodwork (flooring), decorative veneer, furnishing.

PINUS radiata D. Don (RADIATA PINE)
Pin radiata (f) Pino insigne (i) Monterrey Föhre (d) Pino insiginis (e)

Conifer from Southeast Europe, New Zealand, and Australia, in good supply.

Physical properties: density of 31 lbs/cu ft, quite flexible, fairly durable (1.8), straight fibers.

Technical properties: easily glued (permeable), fixed, and screwed. Finish: preliminary sealing required.

Natural durability: low durability (fungi) and sensitive (insects).

Impregnability: impregnable heartwood, fairly pregnable, or fairly impregnable xylem.

Applications: structural plywood boards, exterior (plywood profiles) and interior (paneling), framework, glue laminated timber, and furnishing.

PINUS sylvestris L. (REDWOOD) Pin (f) Pino (i) Föhre weiss (d) Pino silvestre (e)

Conifer from Northern Asia and Europe, in good supply.

Physical properties: density of 31-34 lbs/cu ft, fairly inflexible, fairly durable (2), straight fibers, resin canals.

Technical properties: glued, very resinous, easily nailed, and screwed. Finish: dry well.

Natural durability: low durability (fungi) and sensitive (insects).

Impregnability: impregnable heartwood, pregnable xylem.

Applications: structural plywood boards, interior and exterior woodwork (doors, paneling/cladding, flooring), framework, decorative veneer, furnishing and cabinetmaking, glued laminated timber.

POPULUS alba L. (POPLAR) Peuplier (f) Pioppo (i) Pappel (d) Chopo (e)

Deciduous from Europe, Asia, and Northern Africa, in good supply.

Physical properties: density of 26-30 lbs/cu ft, quite flexible, soft (1.2-2.6), straight fibers.

Technical properties: easily glued, nailed, and screwed. Finish: meshed fibers.

Natural durability: low durability (fungi) and sensitive (insects).

Impregnability: fairly impregnable heartwood, pregnable xylem.

Applications: plywood boards.

PSEUDOTSUGA menziessii Franco (DOUGLAS FIR)

Pin d'Orégon (f) Abete di Douglas (i) Douglasfichte (d) Pino Oregón (e)

Conifer from North America, United Kingdom, and Australia, in good supply.

Physical properties: density of 29-32 lbs/cu ft, quite flexible, fairly durable (2.2), straight fibers.

Technical properties: easily glued (coloration), fixed, and screwed. Finish: preparation.

Natural durability: fairly durable (fungi) and sensitive (insects).

Impregnability: impregnable heartwood, fairly pregnable or fairly impregnable xylem.

Applications: plywood, flooring, interior and exterior woodwork, framework.

QUERCUS robur L. (OAK) Chêne (f) Rovere (i) Stieleiche (d) Roble (e)

Deciduous from Europe, Asia Minor, Northern Africa, in good supply.

Physical properties: density of 42-47 lbs/cu ft, quite flexible, fairly durable (3.5-4.4), straight fibers. Can corrode metal. Bends easily with steam.

Technical properties: easily glued, nailed, and screwed. pre-drilling. Finish: preliminary sealing required.

Natural durability: durable (fungi) and sensitive (insects).

Impregnability: impregnable heartwood, pregnable xylem.

Applications: interior woodwork (flooring), framework, decorative veneer, furnishing, and cabinetmaking.

QUERCUS rubra L. (RED OAK) Chêne rouge (f) Quercia rossa (i) Roteiche (d) Roble rojo (e)

Deciduous from North America, in good supply.

Physical properties: density of 40-49 lbs/cu ft, quite flexible, fairly durable (3.5-4.5), straight fibers. Stains on contact with metal and humidity. Bends easily with steam.

Technical properties: varying ease of gluing, nailing, and screwing: pre-drilling. Finish: preliminary sealing required.

Natural durability: low durability (fungi) and sensitive (insects).

Impregnability: fairly pregnable to fairly impregnable heartwood, pregnable xylem.

Applications: plywood boards, interior woodwork (flooring), decorative veneer, furnishing, and cabinetmaking.

Directory

CASA RAFA 6
Estudi ARCA d'aquitectura
Freneria, 5, pral., 1.ª
08002 Barcelona, Spain
T (34) 933 107 160
F (34) 932 687 018
arc@coac.net

MAISON CONVERCEY 16
Quirot-Vichard Architectes
4, rue de la Prefecture
25000 Besançon, France
T (33) 03 81813131
F (33) 03 81619894
www.quirot-vichard.com

MAISON GOULET 22, 180, 210, 222
Saïa Barbarese Topouzanov Architectes
339 est, rue Saint Paul, Vieux-Montréal
Québec, Canada H2Y 1H3
T (514) 866 2085
F (514) 874 0233
sbt@sbt.qc.ca

CHUN RESIDENCE 28, 184
Chun Studio, Inc.
4040 Del Rey Av., suite 5
Marina del Rey, CA 90292, USA
T 310 821 0415
david@chunstudio.com

ISLAND HOUSE 34, 212, 233
Arkitektstudio Widjedal Racki Bergerhoff
Hornsgatan 79
SE 118 49 Stockholm, Sweden
T (46) 8 556 974 09
F (46) 8 556 974 39
www.wrb.se

SUITCASE HOUSE HOTEL 42, 230
TREASURE PALACE 178, 220, 228
Edge Design Institute Ltd.
Room 1706-08, 663 King's Road
North Point, Hong Kong, China
T 852 2802-6212
F 852 2802-6213
www.edge.hk.com

PUTNEY HOUSE 48
Tonkin Zulaikha Greer Architects
117 Reservoir Street
Surry Hills, NSW 2010 Australia
T 02 9215 4932
F 02 9215 4901
www.tzg.com.au

MAISON PLÉNEUF-VAL-ANDRÉ 54
DL Architects
5, rue Jules Vallees
75011 Paris, France
T (33) 01 53276464
F (33) 01 53276465
www.daufresne-legarrec-architectes.com

HAUS NENNING 60, 246
HAUS HEIN 248
Cuckrowicz-Nachbaur Architekten
Rathausstrasse 2
6900 Bregenz, Austria
T (43) 5574 82788
F (43) 5574 82688
www.cn-arch.at

CASA DMAC 68
CASAS LLIÇÀ/BIGUES 9
Nassar Arquitectes, S.L.
Trafalgar, 17, pral., A
080010 Barcelona, Spain
T (34) 932955800
F (34) 932955801
nassarq@menta.net

DANIELSON HOUSE 74
Brian MacKay-Lyons
2188 Gottingen Street, Halifax
Nova Scotia, Canada B3K 3B4
T 902 429 1867
F 902 429 6276
www.bmlaud.ca

MOUNTVIEW RESIDENCE 80
MATTHEWS RESIDENCE 254
JMA Architects
Level 1, 457 Upper Edward
St. Keynlynn Centre
PO Box 380 Spring Hill, Brisbane
QLD 4004 Australia
T (07) 3839 3794
F (07) 3839 6112
www.jma-arch.com

PILOTPROJECT KFN 86
PASSIVEHAUSANLAGE 158
Johannes Kaufmann, Oskar Leo Kaufmann
Sägerestrasse 4
A-6850 Dornbirn, Austria
T 43 (0) 5572 23690
F 43 (0) 5572 23690-4
www.jkarch.at

CHILMARK RESIDENCE 94
Charles Rose Architects
115 Willow Av.
Somerville, MA 02144, USA
T 617 628 5033
F 617 628 7033
www.charlesrosearchitects.com

VILLA LÅNGBO 100, 182
Olavi Koponen Architect
Apollonleath 23 B 39
00100 Helsinki, Finland
T 43 (0) 5572 23690
F 43 (0) 5572 23690-4
www.kolumbus.fi/olavi.koponen

KOEHLER RETREAT 106
Salmela Architecture & Design
852 Grandview Av.
Duluth, MN 55812, USA
T 218 724 7517
F 218 728 6805
dd.salmela@charter.net

CASA ACAYABA 112, 186
Marcos Acayaba Arquitetos
Rua Helena, 170 – cj. 143
04552 Vila Olímpia São Paulo, SP, Brazil
T 55 11 3845 0738
F 55 11 3849 1045
macayaba@vol.com.br

EAVE HOUSE 120
ROOF HOUSE 174
PORCH HOUSE 242
Tezuka Architects
1-29-2 Tanattumi, Setagaya
Tokyo 158 0087, Japan
T 03 3703 7056
F 03 3703 7038
www.tezuka-arch.com

MEHRFAMILIENHAUS II 126, 200
Vogt Architekten
Rötelstrasse 15
8006 Zürich, Switzerland
T 01 364 3406
F 01 364 3498
www.vogtarchitekten.ch

Directory

GALLOWAY RESIDENCE 132
The Miller-Hull Partnership
Polson Bld., 71 Columbia, sixth floor
Seattle, WA 98104, USA
T 206 682 6837
F 206 682 5692
www.millerhull.com

CASA EN CHAMARTÍN 138
Nieto Sobejano Arquitectos, S.L.
Rodriguez Marín, 61
28016 Madrid, Spain
T (34) 915 643 830
F (34) 915 643 836
nietosobejano@nietosobejano.com

CASA BANDEIRA DE MELLO 146, 188, 232
Mauro Munhoz arquitetura
Av. Brigadeiro Luís Antônio, 4919 Jardim Paulista
São Paulo, SP, Brazil CEP 01401-002
T 55 11 3885 9353
F 55 11 3052 3858
mauro@mauromunhoz.arq.br

STEINWENDTNER HAUS 152
Hertl.Architekten
Zwischenbrücken 4
A-4400 Steyr, Austria
T 43 7252 46944
F 43 7352 47363
www,hertl-arquitekten.com

WOHNHAUS MIT ATELIER 164
Thomas Maurer Architekturbüro
Melchnaustrasse 47G
CH-4900 Langenthal, Switzerland
T 062 922 10 11
F 062 922 10 11
imarch@bluewin.ch

GREAT (BAMBOO) WALL 172, 252
Kengo Kuma & Associates
2-24-8 Minami Aoyama
Minato-ku, Tokyo, 107 0062 Japan
T 81 33401 7121
F 81 33401 7778
www02.so-net.ne.jp/~~kuma/

MICHAELIS HOUSE 176, 208
Harry Levine Architects
19 Boundary Street
Rushcutters Bay, NSW 2011 Australia
T 9380 6033
F 9380 6088
hl@hlarch.au

PINE FOREST CABIN 190
WOOD RESIDENCE 206
Cutler Anderson Architects
135 Parfitt Way SW
Bainbridge Island, WA 98110, USA
T 206 842 4710
F 206 842 4420
www.cutler-anderson.com

RESIDENTIAL IN ÖSCHINGEN 192
WOHNHAUS STUTTGART 194
Schlude Architekten
Kleine Falterstrasse 22
70597 Stuttgart, Germany
T 0711 765 2590
F 0711 765 2592
www.schlude-architekten.de

CHALET PICTET 196, 202, 218, 240
Charles Pictet Architecte
13, rue de Roveray
CH-1207 Geneva, Switzerland
T 206 842 4710
F 206 842 4420
www.pictet-architrcte.ch

DUPLEX RUE DE SAVOIE 198, 216
Littow Architectes
26 bis, rue des Peupliers
92100 Boulogne, France
T 33 (0)1 46 09 00 34
F 33 (0)1 46 09 04 82
littow@magic.fr

LATORRE RESIDENCE 204
Gary Cunningham Architect
Dallas, TX, USA

STUDIO 3773 214
Dry Design
5727 Venice Blv.
Los Angeles, CA 9019, USA
T 1 323 954 9084
F 1 323 954 9085
www.drydesign.com

FLOODED HOUSE 224, 250
HOUSE WITH THREE ERAS 225, 234
GAD Architecture
9 Desbrosses St.
515 New York, NY 10013, USA
T 1 917 345 0368
F 1 212 941 6496
www.gadarchitecture.com

HOUSE R 236
Feyferlik/Fritzer
Glacisstraße 7
A-8010 Graz, Austria
T 0316 34 76 56
F 0316 38 60 29
feyferlik@inode.at fritzer@iniode.cc

AVALON HOUSE 244
Connor + Solomon Architects
Warehouse 5, 37 Nicholson Street
Balmain East, NSW 2041 Australia
T 612 9810 1329
F 612 9810 4109
www.coso.com

PAULMEAD TREEHOUSE 258
Tomas Millar, Tom Howard, James Foottit
Paulmead, Bisley
Gloucestershire, GL6 7AG, UK
tomas@omnivorist.org

LEE TREEHOUSE 264
Joseph Lim
Department of Architecture
National University of Singapore
4 Architecture Drive, Singapore 117566
T 65 68743528
www.josephlimdesigns.com

HARDY HOUSE 270
Cheong Yew Kuan
45, Cantonment Road
Singapore 089748
T 735 5995
F 738 8295
area@indo.net.id

ACKNOWLEDGEMENTS:

To everyone who has contributed – on occasion, beyond what was required of them – in locating and dispatching published material and architectural study documents: owners, architects, designers, photographers, and architecture firm office and documentation personnel. Special thanks go to: Bruno Vincent, Nadia Meratla, Kathy Chu, Christian Oliver, Chad Jamieson, Paolo Denti, Susanne Gmeiner, Melissa Bland, Eri Ishida, and Eduardo Lopes, and in particular to Daniel Nassar, colleague and friend, and master of all things related to wood.

OTHER PHOTOGRAPHY CREDITS

The photographs in the introductory chapter are by the author (p. 67), from Daniel Nassar (p. 8,9,10,11), provided by Holza company (p. 12, 13), and the magnificent nature photographs on pages 1, 5, and 276 are from © Ricardo Vila, also published in *Colores. El paisaje iluminado* (Vila Editorial, 2003).